Chapter 1: Introduction to Oxytocin and Its Role in Human Biology

Oxytocin is often referred to as the "love hormone" or the "bonding hormone" because of its powerful influence on human connection, emotion, and social behavior. However, its role goes far beyond just creating feelings of warmth and affection—it is integral to a wide range of physiological processes that regulate both physical and emotional well-being. As we explore oxytocin in this chapter, we'll uncover its scientific foundation, its diverse functions, and how it shapes human behavior in profound ways. Understanding oxytocin is the first step in mastering the blockers that hinder its effect on our lives.

Overview of Oxytocin as a Hormone and Neurotransmitter

Oxytocin is a peptide hormone produced in the hypothalamus, a region of the brain that plays a central role in regulating many of the body's essential functions. Once synthesized, oxytocin is stored in the posterior pituitary gland and is released into the bloodstream in response to certain stimuli. In addition to its role as a hormone, oxytocin also functions as a neurotransmitter in the brain, where it helps regulate various social and emotional processes.

As a hormone, oxytocin is best known for its role in childbirth and lactation. During labor, it stimulates uterine contractions, helping to facilitate the birth of a child. After birth, oxytocin is crucial in promoting the ejection of milk during breastfeeding, creating a bond between mother and child. These well-known biological functions make oxytocin an essential component of human reproduction.

However, oxytocin's effects extend far beyond childbirth and breastfeeding. As a neurotransmitter, oxytocin influences behavior and emotions, acting on the brain's limbic system, which is responsible for regulating mood, emotions, and social interactions. It plays a critical role in forming and maintaining social bonds, fostering trust, compassion, and empathy in relationships. It also affects mood regulation, making it an important factor in mental health and well-being.

Functions in Bonding, Childbirth, and Social Behavior

One of oxytocin's most well-recognized functions is its role in bonding, particularly in the context of parent-child relationships. From the moment of birth, oxytocin is involved in creating a deep emotional connection between a mother and her newborn. This bonding is not only vital for the survival of the infant but also for the emotional development of the child. Oxytocin promotes caregiving behaviors in parents, encouraging nurturing and protective instincts.

The effects of oxytocin in bonding extend beyond the parent-child relationship. This hormone is involved in romantic relationships, where it helps foster intimacy, trust, and attachment between partners. Oxytocin's role in creating these strong emotional connections also extends to friendships and social networks, where it strengthens cooperation and empathy between individuals. The hormone's influence is also seen in group dynamics, helping to build a sense of belonging and support within communities.

Research has shown that oxytocin levels are higher in individuals who engage in acts of kindness, empathy, and social connection. Even simple gestures, like hugging or comforting someone in distress, can stimulate the release of oxytocin, creating a positive feedback loop that enhances feelings of well-being and social cohesion.

The Significance of Oxytocin in Physical and Emotional Well-Being

Oxytocin plays an essential role in both physical and emotional health. It has been shown to have several therapeutic benefits, including reducing stress, lowering blood pressure, and enhancing immune function. By reducing the production of stress hormones like cortisol, oxytocin helps to maintain a sense of calm and relaxation, even in the face of challenging situations.

Emotionally, oxytocin has been linked to feelings of happiness, safety, and well-being. It is often associated with the "feel-good" emotions we experience when we connect with others in meaningful ways. This can include the joy of being with loved ones, the comfort of friendship, or the satisfaction of helping others. In fact, research has found that higher levels of oxytocin are associated with lower levels of anxiety and depression, making it a key player in emotional regulation.

Furthermore, oxytocin can promote healing after trauma. Studies suggest that oxytocin's calming and nurturing effects can help individuals recover from emotional wounds, especially those caused by loss or social isolation. It enhances our ability to form trusting relationships and helps us maintain emotional resilience in the face of life's difficulties.

Understanding Oxytocin Blockers

Despite the many positive effects of oxytocin, there are factors that can block or hinder its release and activity. In the following chapters, we will explore what these blockers are, how they impact our behavior, and why they are an important aspect of mastering oxytocin in our lives. Whether through external factors like stress or internal physiological issues, these blockers can disrupt the delicate balance oxytocin provides, leading to a range of emotional and social challenges.

As we delve deeper into the topic of oxytocin blockers, it's important to understand the role oxytocin plays in facilitating connection and well-being. Without a proper understanding of this hormone's power, it's difficult to recognize and overcome the obstacles that hinder its optimal function.

Chapter 2: The Basics of Oxytocin Blockers

In Chapter 1, we explored the vital role of oxytocin in human biology, its profound impact on bonding, social behavior, and emotional well-being. However, the powerful effects of oxytocin can be counteracted by substances and conditions known as *oxytocin blockers*. In this chapter, we will define what these blockers are, how they affect the body and brain, and explain the complex relationship between oxytocin and its blockers. Understanding these blockers is crucial for unlocking the full potential of oxytocin and achieving better connection, health, and well-being.

What Are Oxytocin Blockers?

Oxytocin blockers are factors—whether biological, environmental, or psychological—that interfere with the natural release or action of oxytocin in the body. They inhibit the hormone's ability to promote bonding, reduce stress, and foster feelings of empathy and compassion.

The body's natural processes and external environmental stressors can block oxytocin in various ways. These blockers can be temporary, such as those caused by acute stress, or they can be chronic, arising from long-term trauma, illness, or negative social environments. Oxytocin blockers can also be pharmacological, resulting from medications or substances that interfere with the body's natural hormonal regulation.

At the core, these blockers work by either reducing the synthesis of oxytocin, interfering with its release, or blocking its receptors in the brain and body. Without functioning oxytocin pathways, many of the positive effects the hormone provides—emotional regulation, social bonding, stress relief—become impaired.

How Do Oxytocin Blockers Affect the Body and Brain?

Oxytocin's role in the body is multifaceted, and its blockers have wide-ranging effects. When oxytocin release is blocked, individuals may experience a cascade of physical, emotional, and psychological symptoms that can affect social behavior, health, and overall well-being.

1. **Disruption of Social Bonds**

   Oxytocin is known as the "social glue" of human interaction. Its absence or blockade in social contexts can create difficulties in forming and maintaining connections with others. People may become less empathetic, less trusting, and less likely to seek out or provide social support. The emotional connection between partners, friends, and family members may weaken, leading to feelings of loneliness and isolation. For example, a person who experiences chronic stress or trauma may find it difficult to connect with loved ones, as the body's ability to release oxytocin is compromised.

2. **Increased Stress and Anxiety**

   Oxytocin is a potent counter to the body's stress response, helping to lower levels of cortisol—the hormone responsible for feelings of stress. When oxytocin production is blocked, stress levels can increase, resulting in higher levels of anxiety, depression, and emotional instability. In situations where oxytocin would normally promote calmness and reassurance (such as through touch, bonding, or caregiving), the absence of this hormone can leave a person feeling on edge, hypervigilant, or emotionally distant.

3. **Impaired Emotional Regulation**

   The presence of oxytocin allows for smoother emotional regulation, as it helps individuals process their emotions and respond to social situations in a balanced way. Oxytocin blockers, on the other hand, can make it difficult to manage emotional states. This can result in emotional outbursts, difficulty calming down after stressful events, and an overall sense of emotional numbness. In extreme cases, it may even contribute to emotional disorders such as depression, post-traumatic stress disorder (PTSD), or personality disorders.

4. **Weakened Immune Response**

   Beyond emotional and social implications, oxytocin also plays a crucial role in physical health by supporting the immune system. Research has shown that higher oxytocin levels can enhance immune function, making the body more resistant to illness and aiding in faster recovery after injury. Conversely, when oxytocin levels are blocked, immune function can become compromised, potentially leading to a greater susceptibility to infections and slower recovery from illness.

5. **Reduced Pain Tolerance**

Oxytocin is involved in the modulation of pain. It helps lower pain sensitivity, particularly during times of stress or emotional distress. The blockade of oxytocin can result in an increased perception of pain, whether physical or emotional, making individuals less resilient to physical injuries or the emotional turmoil that can come with difficult life experiences.

The Relationship Between Oxytocin and Its Blockers

Understanding how oxytocin and its blockers interact is central to mastering their effects on our lives. Oxytocin blockers do not just negate the positive effects of oxytocin—they also create a feedback loop where the absence of oxytocin leads to a cascade of negative outcomes. For example, as we saw with stress and anxiety, the lack of oxytocin can increase cortisol production, which in turn reduces oxytocin even further. This loop perpetuates negative feelings, making it harder for individuals to break free from the cycle of emotional and physical dysfunction.

There is also a physiological relationship between oxytocin and other neurochemicals. For example, substances like dopamine and serotonin, which are responsible for regulating mood, are influenced by oxytocin levels. When oxytocin is blocked, these neurochemicals may be less effective in providing emotional balance, leading to feelings of sadness, frustration, or disconnection.

It's also important to recognize that oxytocin blockers are not one-size-fits-all. The nature and strength of the blockade depend on the individual's unique biology, life experiences, and environmental factors. For instance, chronic stress or unresolved trauma may create a more entrenched oxytocin blockade than occasional emotional distress, and different people may respond to similar stressors in different ways.

Types of Oxytocin Blockers

Oxytocin blockers can come in many forms, from natural, internal blockages to external, pharmacological agents. Some of the most common categories include:

1. **Psychological Blockers**

   Negative emotions such as fear, anger, or distrust can block oxytocin release. Trauma, chronic stress, unresolved emotional conflict, or negative past experiences can all create a psychological block to oxytocin, making it harder to form new bonds or heal existing ones.

2. **Physiological Blockers**

   Medical conditions that affect the brain or endocrine system—such as certain neurological disorders, hormonal imbalances, or chronic diseases—can impair oxytocin production. For example, some forms of depression and anxiety have been linked to reduced oxytocin activity in the brain.

3. **Social and Environmental Blockers**

   Isolation, lack of social support, or toxic environments can serve as external blockers to oxytocin. Living in a high-stress urban environment, having a stressful job, or engaging in unhealthy social interactions can all lead to a reduction in oxytocin levels, leaving individuals feeling disconnected and emotionally distant.

4. Pharmacological Blockers

Certain medications and substances can directly interfere with oxytocin. For example, certain anti-anxiety medications, antidepressants, or other psychiatric drugs may alter oxytocin pathways, while chronic alcohol or drug use can have similar effects. Additionally, synthetic blockers may be prescribed in certain medical treatments to achieve specific physiological goals, although these come with their own set of risks and ethical concerns.

Conclusion

The concept of oxytocin blockers is crucial for understanding the delicate balance that oxytocin creates in human biology and behavior. These blockers can have profound effects on an individual's ability to bond, regulate emotions, and maintain social and physical health. By recognizing how these blockers manifest—whether through psychological, physiological, environmental, or pharmacological factors—we can begin to develop strategies to overcome them, restore oxytocin levels, and create a more connected, fulfilling life. In the next chapters, we will dive deeper into the neurochemistry behind oxytocin and its blockers, exploring how these blockers impact the brain and the pathways that control our emotions and behaviors.

Chapter 3: Neurochemical Pathways and Oxytocin Receptors

Understanding the molecular mechanisms of oxytocin is critical to mastering its blockers. While we have explored what oxytocin is and how its blockers disrupt social, emotional, and physical health, this chapter will take a deeper dive into the neurochemistry of oxytocin. Specifically, we will explore the neurochemical pathways it operates through, the role of oxytocin receptors in the body and brain, and how blockers interfere with these pathways to inhibit oxytocin's beneficial effects.

The Neurochemistry of Oxytocin

Oxytocin is a complex molecule that acts both as a hormone and a neurotransmitter, influencing a wide range of behaviors and physiological processes. It is synthesized primarily in the hypothalamus, a region of the brain responsible for regulating many of the body's essential functions, including hunger, thirst, and body temperature. From the hypothalamus, oxytocin is transported to the posterior pituitary gland, where it is released into the bloodstream. Once in circulation, oxytocin reaches various organs and tissues, where it influences everything from uterine contractions during childbirth to milk ejection in breastfeeding mothers.

But oxytocin's influence extends far beyond its role in reproduction. In the brain, oxytocin is a key player in regulating social behaviors. It promotes trust, empathy, and bonding between individuals. It is known as the "bonding hormone" because it helps facilitate the formation of relationships, whether between parents and children, romantic partners, or friends. Oxytocin also plays a role in reducing stress, boosting mood, and increasing feelings of well-being.

While oxytocin's presence in the bloodstream is important, its actions in the brain are mediated by a network of neurochemical pathways, which help it communicate with specific regions of the brain involved in social cognition, emotional regulation, and memory.

Oxytocin Receptors in the Brain and Body

For oxytocin to exert its effects, it must bind to specific receptors located throughout the body. These receptors are proteins embedded in the cell membranes of target cells, and when oxytocin binds to them, it triggers a cascade of biochemical events inside the cell. Oxytocin receptors are found in many areas of the brain, including regions associated with emotion, social behavior, and memory, such as the amygdala, the prefrontal cortex, and the hippocampus.

1. **Amygdala**

   The amygdala is the brain's emotional center, responsible for processing fear, pleasure, and other intense emotional reactions. Oxytocin acts as a modulator in the amygdala, reducing the effects of stress and anxiety. It does this by decreasing the amygdala's sensitivity to negative stimuli, thereby facilitating more positive social interactions and reducing emotional reactivity. This is why oxytocin is often referred to as a calming hormone. Blockers that interfere with oxytocin in this region can leave individuals feeling more anxious, fearful, or distrustful.

2. **Prefrontal Cortex**

   The prefrontal cortex is crucial for decision-making, impulse control, and social behavior. Oxytocin receptors in this area are thought to play a significant role in promoting prosocial behaviors, such as empathy, cooperation, and moral reasoning. When oxytocin binds to receptors in the prefrontal cortex, it enhances our ability to connect with others, understand their perspectives, and make decisions that prioritize others' needs. Blockers that interfere with this process can impair empathy, emotional intelligence, and interpersonal communication.

3. **Hippocampus**

   The hippocampus is primarily involved in memory formation and spatial navigation, but it also has a role in emotional regulation. Oxytocin receptors in the hippocampus help moderate emotional memory, promoting positive associations with others and facilitating the formation of long-term bonds. By strengthening the emotional memory of positive experiences, oxytocin helps reinforce trust and social connection. Blockers that interfere with this region can impair the ability to form trusting relationships or recall positive experiences, making it more difficult to build and maintain social connections.

4. **Peripheral Oxytocin Receptors**

   Outside of the brain, oxytocin receptors are found in various tissues throughout the body, including the uterus, mammary glands, and cardiovascular system. In these areas, oxytocin plays a role in smooth muscle contraction (e.g., during childbirth), milk production, and even the regulation of blood pressure. Oxytocin's ability to promote relaxation and homeostasis in the body can be disrupted if its receptors are blocked, leading to increased stress or physical discomfort.

How Blockers Interfere with Oxytocin Receptors

Oxytocin blockers can interfere with oxytocin's ability to bind to its receptors, thereby preventing it from exerting its usual effects on the brain and body. These blockers can be either **biological** (internal) or **pharmacological** (external), and their impact on oxytocin receptors varies depending on the type of blocker involved.

Biological Blockers

For example, individuals who have suffered from childhood neglect or emotional abuse may have fewer oxytocin receptors in key areas of the brain, making it more difficult for them to form secure attachments with others. This is often referred to as "attachment trauma," and it can have long-lasting effects on relationships and emotional well-being.

Pharmacological Blockers

antidepressants

Antipsychotic drugs

alcohol

drugs

In addition to prescription medications, some pharmaceutical interventions are specifically designed to block oxytocin as a way of controlling certain physiological processes. For example, **antagonists of oxytocin receptors** may be used in specific medical treatments to manage premature labor or to prevent excessive uterine contractions during childbirth. However, while these drugs are effective in their targeted applications, they can have unintended side effects on emotional bonding and social behavior.

Social and Environmental Blockers

chronic loneliness

toxic relationships

social isolation

The Impact of Blocked Oxytocin Receptors

The consequences of blocked oxytocin receptors are far-reaching. Oxytocin's blockade in the brain impairs not only emotional regulation but also social cognition and bonding. People with reduced oxytocin receptor sensitivity may struggle with establishing and maintaining meaningful relationships. They may experience increased difficulty with emotional regulation, heightened stress responses, and decreased social engagement.

The blockade of oxytocin receptors can also contribute to a wide range of psychological disorders, including **anxiety**, **depression**, and **post-traumatic stress disorder (PTSD)**. When oxytocin's calming and bonding effects are blocked, individuals may feel more isolated, less able to cope with stress, and less motivated to form or maintain close relationships. Furthermore, they may struggle with feelings of distrust, emotional numbness, and difficulty in understanding or expressing emotions.

Conclusion

Oxytocin's effects on human behavior and physiology are deeply rooted in its ability to bind to specific receptors in the brain and body. Understanding these receptors—and how they interact with blockers—is critical for addressing the social and emotional challenges associated with oxytocin imbalance. In the next chapters, we will explore how natural and synthetic blockers affect these pathways and how we can actively work to restore oxytocin balance and improve emotional and social well-being. By mastering these neurochemical pathways, we can unlock the full potential of oxytocin in enhancing our connections with others and ourselves.

Chapter 5: Synthetic Oxytocin Blockers

In addition to the natural mechanisms that can block oxytocin, there are also synthetic blockers that can inhibit oxytocin's effects. These pharmaceutical agents can have profound impacts on the body's ability to regulate oxytocin levels, potentially interfering with emotional bonding, social behaviors, and physiological processes like childbirth and lactation. Understanding how these synthetic blockers work, their medical uses, and the ethical implications surrounding their use is crucial for mastering oxytocin's role in human behavior and biology.

Pharmaceutical Blockers: How They Work

Synthetic oxytocin blockers, also known as **oxytocin antagonists**, are compounds that bind to oxytocin receptors and prevent the natural hormone from exerting its effects. Oxytocin antagonists can block both the peripheral and central actions of oxytocin, preventing the hormone from influencing the body's smooth muscles, reproductive organs, and brain regions responsible for emotional regulation and social bonding.

Pharmaceutical oxytocin blockers are typically used in clinical settings to manage specific health conditions, particularly those related to childbirth or certain mental health disorders. The blockers can either be injected directly into the bloodstream or delivered through oral medications, depending on the condition being treated and the specific blocker being used.

While there are several drugs that can inhibit oxytocin's effects, the most commonly used ones include:

1. **Atosiban** – A competitive oxytocin receptor antagonist used in obstetrics to prevent premature labor. It works by blocking oxytocin's uterine-stimulating effects, helping to prevent preterm contractions and delays the onset of labor.

2. **Naloxone** – While primarily known as an opioid antagonist, naloxone has also been shown to interfere with oxytocin signaling in some studies. It can temporarily block oxytocin's effects on the brain, especially in the context of substance use disorder treatment, where oxytocin's calming effects may interact with the addiction pathways.

3. **Antipsychotic medications** – Some psychiatric drugs, particularly those targeting the dopamine system, can indirectly affect oxytocin signaling. These medications may inhibit oxytocin release or reduce receptor sensitivity, contributing to the emotional and social disconnection often observed in patients taking these drugs.

Medical Uses of Oxytocin Blockers

While oxytocin blockers can interfere with normal physiological and emotional processes, they also have essential medical applications in specific situations. Below are some of the key medical uses of synthetic oxytocin blockers:

1. Preventing Preterm Labor

One of the primary uses of oxytocin blockers is in the treatment of preterm labor. During pregnancy, the body's natural release of oxytocin triggers uterine contractions, which can lead to premature birth if they occur too early. Oxytocin antagonists like **atosiban** are used in obstetrics to slow down or prevent premature labor by inhibiting oxytocin's uterine-stimulating effects. This allows doctors to delay delivery, giving the baby more time to develop before birth.

However, the use of these drugs is not without controversy. Some research suggests that long-term exposure to oxytocin blockers could interfere with normal bonding processes between the mother and infant, leading to challenges in breastfeeding and maternal attachment.

2. Managing Postpartum Hemorrhage

Oxytocin plays a vital role in controlling uterine bleeding after childbirth by stimulating contractions that help the uterus return to its pre-pregnancy size. In cases where oxytocin is needed to reduce excessive postpartum bleeding, synthetic blockers may be administered to control unwanted uterine contractions or to counteract overstimulation.

Although oxytocin blockers can be used effectively in this context, their overuse or misuse may have unintended effects on the mother's ability to form a secure bond with her newborn. As such, careful monitoring is required when using oxytocin blockers during labor and delivery.

3. Treating Autism Spectrum Disorders (ASD) and Other Psychiatric Conditions

Oxytocin has gained attention in recent years for its potential role in treating various psychiatric conditions, particularly those that affect social bonding and emotional regulation. Some research suggests that individuals with **autism spectrum disorder (ASD)** have lower oxytocin levels and may benefit from treatments that enhance oxytocin signaling. However, in some cases, synthetic oxytocin blockers may be used in combination with other treatments to manage behaviors associated with these disorders.

For instance, **antipsychotic medications**, which may have a secondary effect of blocking oxytocin activity, are sometimes prescribed for individuals with ASD, particularly to address aggression or irritability. The downside is that long-term use of such medications can disrupt social interactions and emotional engagement, often leading to a sense of detachment or emotional numbness.

4. Managing Substance Abuse and Addiction

Oxytocin plays a role in the brain's reward system and has been implicated in both addiction and the recovery process. The hormone is thought to promote bonding and attachment, and lower levels of oxytocin may contribute to increased susceptibility to addiction. Some studies suggest that **oxytocin antagonists** could be used to manage the addictive behaviors associated with drugs like opioids or alcohol by blocking oxytocin's reinforcing effects.

However, the use of oxytocin blockers in addiction treatment remains controversial. While they may reduce some addictive behaviors in the short term, blocking oxytocin could also exacerbate the social isolation and emotional numbness often experienced by those with substance use disorders. The long-term effects of using oxytocin blockers in addiction treatment are not fully understood and require further research.

Ethical Considerations

The use of synthetic oxytocin blockers raises important ethical questions, particularly when it comes to long-term consequences on emotional well-being and social behaviors. While oxytocin blockers can have beneficial uses in specific medical contexts, their potential to interfere with human connection and emotional regulation cannot be overlooked.

- **Informed Consent**: Patients receiving oxytocin blockers, particularly during labor or in psychiatric treatment, must be fully informed of the potential risks and side effects. This includes understanding how the drug might affect their ability to bond with their baby, experience empathy, or maintain emotional regulation.

- **Impact on Attachment**: The use of oxytocin blockers in labor and delivery raises concerns about their impact on the mother-child bond. Oxytocin's role in maternal bonding is well-documented, and disrupting this process may have lasting effects on the parent-child relationship.

- **Psychiatric Implications**: For individuals with mental health disorders, the use of synthetic blockers may present a trade-off between managing symptoms and potentially hindering social and emotional development. In particular, those with depression, anxiety, or autism may experience a reduction in emotional responsiveness and empathy, making it harder to engage in positive social interactions.

- **Addiction Treatment**: The application of oxytocin blockers in addiction treatment must be considered with caution. While oxytocin may play a role in reinforcing addictive behaviors, blocking its effects could interfere with the emotional healing process necessary for long-term recovery. The complex interplay between oxytocin and addiction pathways means that oxytocin antagonists should be used judiciously and as part of a comprehensive treatment approach.

Potential Side Effects and Risks

Like all medications, synthetic oxytocin blockers come with a set of potential risks and side effects. These include:

- **Emotional Numbness**: One of the most concerning side effects is emotional numbness, where individuals may experience difficulty in forming or maintaining relationships, reduced empathy, and a diminished sense of emotional connection.
- **Increased Stress**: Oxytocin helps regulate the body's stress response, and blocking it could result in an increase in cortisol (the stress hormone), potentially exacerbating anxiety, irritability, or depressive symptoms.
- **Disrupted Bonding**: As mentioned, the blockade of oxytocin can interfere with bonding, especially in the context of childbirth or parenting. Mothers may struggle to bond with their infants, which can have long-term developmental implications for the child.
- **Physical Side Effects**: In the case of obstetric applications, synthetic blockers like atosiban may cause side effects such as nausea, headaches, or dizziness. More serious complications, though rare, may include cardiovascular issues or allergic reactions.

Conclusion

While synthetic oxytocin blockers are crucial tools in managing certain medical conditions, their impact on human connection and emotional regulation must be carefully considered. These blockers offer valuable benefits in specific contexts, such as managing preterm labor, controlling postpartum bleeding, or addressing certain psychiatric conditions. However, the potential long-term consequences of blocking oxytocin—especially on attachment, bonding, and social behavior—should not be underestimated.

As we continue to explore the broader role of oxytocin in human connection, it is essential to strike a balance between the therapeutic uses of synthetic blockers and the need to preserve the natural biological processes that foster emotional health and social well-being.

Chapter 6: The Impact of Oxytocin Blockers on Social Interaction

Oxytocin, often referred to as the "love hormone" or "bonding hormone," plays an essential role in how humans interact with one another. It is involved in promoting trust, social bonding, empathy, and cooperative behaviors. When oxytocin's effects are blocked—either naturally through stress, fear, or trauma, or synthetically through medication—these social functions can become impaired, leading to significant changes in how individuals relate to each other. This chapter delves into the effects of oxytocin blockers on human interaction, empathy, trust, and emotional bonding, as well as real-life examples of how these disruptions manifest in relationships.

The Role of Oxytocin in Social Interaction

Before exploring the effects of its blockers, it's important to understand the fundamental role oxytocin plays in human connection:

1. **Facilitating Social Bonding**: Oxytocin is crucial in forming deep, lasting connections with others. It is released during moments of physical touch, social interactions, and even in moments of shared emotional experiences. Oxytocin enhances the feelings of trust and affection, promoting positive social engagement.

2. **Promoting Empathy**: High levels of oxytocin are linked to an increased ability to understand others' emotions. It helps individuals feel more attuned to the feelings and needs of others, facilitating cooperation, caregiving, and nurturing behaviors.

3. **Building Trust**: Oxytocin has a direct role in promoting interpersonal trust. It strengthens the belief that others will act in one's best interest, which is vital for forming strong, supportive relationships—whether familial, romantic, or professional.

4. **Reducing Social Anxiety**: For those struggling with social anxiety, oxytocin can help lower the threshold for fear in social settings, making it easier to approach others, engage in conversations, and form new relationships.

When oxytocin is blocked or its effects diminished, the mechanisms that support these fundamental aspects of social interaction are compromised, often leading to significant challenges in personal and professional relationships.

How Oxytocin Blockers Alter Human Interaction

Oxytocin blockers—whether they arise naturally through stress or trauma or are introduced through pharmaceutical interventions—can have wide-ranging effects on an individual's social behavior. Here are the primary ways in which oxytocin blockers disrupt human interaction:

1. Decreased Empathy and Emotional Sensitivity

Without adequate oxytocin, individuals may find it more difficult to empathize with others. Emotional responses to the feelings of others may be dulled or less authentic, making it harder to form close relationships. People may appear more indifferent or disconnected, even in emotionally charged situations.

Real-Life Example

2. Impaired Trust and Attachment

Oxytocin plays a critical role in the formation of trust. When oxytocin blockers interfere with this process, individuals may become more suspicious, guarded, or unwilling to open up to others. This can lead to a breakdown in relationships, both romantic and platonic.

Real-Life Example

3. Social Withdrawal and Isolation

Oxytocin's effects extend beyond emotional understanding—its presence encourages social engagement and the desire for companionship. When oxytocin production is blocked, individuals may feel more isolated or reluctant to engage with others. This can lead to avoidance behaviors, where individuals intentionally distance themselves from social opportunities to protect themselves from potential emotional distress.

Real-Life Example

4. Difficulty in Rebuilding Relationships After Conflict

Oxytocin is key in the reconciliation process after conflicts or misunderstandings. It facilitates emotional healing, enabling individuals to move past disagreements, forgive, and reestablish trust. When oxytocin's effects are blocked, individuals may struggle to let go of grudges, making it difficult to repair damaged relationships.

Real-Life Example

5. Impaired Parenting and Family Dynamics

Oxytocin is especially important in the early stages of parenting, as it helps establish the crucial bond between a mother (or primary caregiver) and her newborn. This bond is foundational for the child's emotional development. Blockers—whether caused by stress, trauma, or medications—can disrupt this early attachment, potentially affecting the child's ability to trust and bond with others later in life.

Real-Life Example

The Effect on Relationships, Trust, and Emotional Bonding

When oxytocin blockers impact social interaction, they also alter the foundational components of human relationships: trust, emotional intimacy, and mutual respect. Here are the key effects on these essential relationship dynamics:

1. **Relationships**: The quality of relationships depends heavily on the ability to trust, empathize, and feel emotionally safe. Without the soothing influence of oxytocin, relationships can become tense, emotionally distant, or even volatile. Partners may feel less connected or unable to provide the support the other person needs.

2. **Trust**: Trust is the cornerstone of any meaningful relationship, whether romantic, familial, or professional. Oxytocin blockers can make it difficult for individuals to believe in others' good intentions, causing them to retreat into protective emotional shells. This affects everything from friendships to workplace dynamics, where collaboration and cooperation rely on mutual trust.

3. **Emotional Bonding**: Emotional bonding is essential in forming lasting connections. Oxytocin facilitates this bonding by promoting feelings of warmth, affection, and security. Blockers disrupt this process, leading to weaker bonds, increased emotional distance, and difficulties in developing long-term, stable relationships.

Real–Life Examples of Relationships Affected by Oxytocin Blockers

Case 1: Romantic Relationships

Jenna and Tom had been in a relationship for three years when Tom's behavior began to shift. He became distant, less affectionate, and frequently withdrew from physical touch. Jenna felt that something had changed, but despite her best efforts to communicate, Tom seemed emotionally unavailable.

Tom had been under immense stress at work, and the constant pressure led to a significant decrease in his oxytocin levels. His inability to bond with Jenna created a rift in their relationship. Over time, Jenna began to question the stability of their future together.

It wasn't until Tom sought therapy and explored the underlying effects of his stress that he recognized how the reduction in oxytocin had affected his ability to connect emotionally with Jenna. With targeted stress management and mindfulness techniques, he was able to restore the emotional connection, and the couple began working together to rebuild trust.

Case 2: Parental Bonding

Sarah had always dreamed of becoming a mother, but after the birth of her first child, she struggled with feelings of isolation and disconnection. Despite her overwhelming love for her baby, she felt emotionally numb and found it hard to bond. This made caring for her newborn feel like an exhausting chore rather than a rewarding experience.

Sarah's healthcare provider diagnosed her with postpartum depression, which was inhibiting her natural oxytocin release. By engaging in therapy, stress reduction techniques, and a support system, Sarah gradually began to experience the nurturing bond with her baby that is typically formed in the early days of motherhood.

Conclusion

Oxytocin blockers—whether stemming from natural internal processes or external factors like medications—significantly impact the way humans interact, bond, and communicate. From decreased empathy to impaired trust and social withdrawal, these blockers can have profound effects on relationships. However, understanding how these blockers manifest can help individuals and communities develop strategies to overcome their disruptive effects, restoring emotional connections, and fostering stronger, more empathetic relationships. Recognizing and addressing oxytocin blockers in all forms is the first step toward creating a more connected and emotionally supportive world.

Chapter 7: Oxytocin Blockers and Mental Health

Oxytocin, the "bonding hormone," is deeply entwined with emotional regulation and social behaviors. It is crucial for promoting positive mental health by fostering connection, trust, and emotional balance. However, when oxytocin levels are disrupted or blocked, a cascade of negative effects can unfold, exacerbating existing mental health conditions such as anxiety, depression, and PTSD. This chapter explores the role of oxytocin in mental health, how blockers intensify mental health challenges, and strategies for overcoming the impact of oxytocin disruptions.

The Role of Oxytocin in Mental Health

Oxytocin plays a pivotal role in maintaining emotional well-being. It influences the brain's reward systems, helps regulate mood, and is a key factor in emotional resilience. Here are the primary ways oxytocin supports mental health:

1. **Regulating Stress**: Oxytocin works to counterbalance the effects of stress by reducing the production of cortisol, the primary stress hormone. It helps promote feelings of calm and emotional stability, even in challenging situations.

2. **Enhancing Social Connectivity**: High levels of oxytocin are linked to improved social interactions, which in turn support mental health. Positive social bonds are critical in preventing isolation, loneliness, and depression.

3. **Promoting Positive Emotional Responses**: Oxytocin's influence on the brain increases feelings of happiness, contentment, and well-being. It fosters a sense of security and emotional warmth in relationships, reducing the likelihood of mental health disturbances.

4. **Reducing Anxiety and Depression**: Oxytocin can lower anxiety levels by promoting trust and reducing feelings of fear. Additionally, oxytocin's connection with serotonin and dopamine pathways helps stabilize mood and alleviate depression.

How Oxytocin Blockers Exacerbate Mental Health Conditions

When oxytocin is blocked or its effects suppressed, it disrupts these essential mental health functions. The effects of oxytocin blockers on mental health can be profound, leading to heightened emotional distress, social withdrawal, and an increased risk of developing or worsening mental health conditions. Here's how oxytocin blockers can exacerbate some of the most common mental health challenges:

1. Anxiety

Oxytocin plays an important role in reducing feelings of anxiety by enhancing trust and promoting calm in social situations. When oxytocin is blocked, the body's ability to regulate fear responses is compromised. Individuals may become more easily overwhelmed by anxiety, feeling isolated and unable to connect with others in supportive ways.

- **How It Happens**: Without oxytocin's calming influence, the body remains in a heightened state of stress and vigilance. This leads to overactive fear responses in social situations, increasing anxiety about interactions and perpetuating cycles of stress.
- **Real-Life Example**: Sarah, a young professional, began experiencing severe social anxiety after a series of stressful life events. Her anxiety worsened over time, especially in group settings, where she felt disconnected from others. Therapy revealed that a combination of trauma and prolonged stress had blocked her natural oxytocin release, making her feel increasingly isolated and distrustful of others.

2. Depression

Oxytocin's effects on mood regulation are well-documented. When oxytocin levels drop, individuals may feel more emotionally numb or disconnected. Lack of bonding, social withdrawal, and feelings of isolation can contribute significantly to depression.

- **How It Happens**: Without the emotional uplift that oxytocin provides, individuals can spiral into feelings of loneliness, worthlessness, and hopelessness. Oxytocin's role in the reward system is critical; without it, dopamine levels decrease, making it difficult to experience joy or satisfaction.
- **Real-Life Example**: John, who had struggled with depression for years, found that even when he was surrounded by friends and family, he felt emotionally distant. He learned that his low oxytocin levels were making it difficult to bond with others, which perpetuated his feelings of isolation and sadness.

3. Post-Traumatic Stress Disorder (PTSD)

Oxytocin plays a healing role in trauma recovery, as it helps to regulate the body's stress response and encourages the rebuilding of trust in others. For individuals suffering from PTSD, the presence of oxytocin can reduce the physiological and emotional effects of traumatic memories.

- **How It Happens**: In trauma survivors, oxytocin blockers can make it more difficult to manage intrusive thoughts, hypervigilance, and emotional numbing. The lack of oxytocin's soothing effects on the brain leaves the individual in a heightened state of alert, preventing emotional healing.

- **Real-Life Example**: Mark, a combat veteran, found that his PTSD symptoms intensified after his experiences in the military. Despite attending therapy, he felt emotionally detached and unable to reconnect with his loved ones. His lack of oxytocin was preventing him from establishing the connections needed for healing.

4. Social Isolation and Loneliness

One of the most devastating consequences of oxytocin blockers is a heightened sense of social isolation. Oxytocin promotes social bonding and helps individuals feel connected to others. When oxytocin is blocked, individuals may withdraw from social interactions, feeling as though they do not belong or cannot trust others.

- **How It Happens**: Oxytocin's absence leaves individuals more likely to perceive social situations as threatening or overwhelming, leading them to avoid others. This avoidance leads to loneliness, which exacerbates feelings of depression and anxiety.

- **Real-Life Example**: Lily, a college student, began withdrawing from friends and family after a difficult breakup. Despite her deep desire to connect with others, she felt emotionally distant and unable to form meaningful bonds. Therapy uncovered that her trauma had suppressed her oxytocin levels, making social interactions seem emotionally taxing.

Strategies for Overcoming the Impact of Oxytocin Blockers

While the effects of oxytocin blockers on mental health can be profound, there are several strategies that can help individuals restore balance and improve their emotional well-being. These strategies target both the reduction of blockers and the promotion of natural oxytocin production.

1. Stress Management

Chronic stress is a significant natural blocker of oxytocin. By managing stress, individuals can reduce cortisol levels and enhance their body's ability to produce oxytocin. Stress-reduction techniques such as mindfulness meditation, deep breathing, and progressive muscle relaxation can help create a more supportive internal environment for oxytocin production.

Actionable Tip

2. Therapeutic Approaches

Psychotherapy, particularly trauma-focused therapy, can help individuals process past traumas and rebuild their ability to trust and bond with others. Therapies like Cognitive Behavioral Therapy (CBT), Eye Movement Desensitization and Reprocessing (EMDR), and Emotion-Focused Therapy (EFT) can help address emotional blockages caused by trauma and stress, fostering the restoration of healthy oxytocin levels.

Actionable Tip

3. Building Healthy Relationships

Oxytocin is naturally released during positive social interactions, including bonding with family, friends, and romantic partners. Building and maintaining healthy relationships can be an effective way to naturally increase oxytocin levels. Activities like hugging, laughing together, and engaging in acts of kindness all promote oxytocin release.

Actionable Tip

4. Physical Activity

Exercise is one of the most powerful ways to boost oxytocin levels. Physical activity promotes the release of endorphins and oxytocin, helping to alleviate feelings of depression and anxiety. Regular exercise not only improves mood but also enhances emotional resilience by strengthening the body's ability to manage stress.

Actionable Tip

5. Nutritional Support

Certain foods and supplements have been shown to support oxytocin production. Omega-3 fatty acids, magnesium, and vitamin D are known to positively affect mood and mental well-being, while foods rich in antioxidants, such as berries, nuts, and leafy greens, can help reduce stress and inflammation in the body.

Actionable Tip

Conclusion

Oxytocin is a powerful tool for maintaining mental health. When oxytocin levels are blocked—whether due to stress, trauma, or other factors—mental health conditions such as anxiety, depression, and PTSD can become more pronounced. However, by recognizing the role of oxytocin in emotional regulation and implementing strategies to overcome the blockers, individuals can begin to heal and restore balance to their mental well-being. By focusing on stress reduction, therapy, healthy relationships, exercise, and nutrition, we can unlock the full potential of oxytocin and enhance our ability to cope with life's challenges.

Chapter 8: Blocking Oxytocin in Modern Society

In today's fast-paced, technology-driven world, the natural processes that once nurtured human connection and emotional bonding are increasingly being disrupted. The modern environment—marked by technology, social media, and urbanization—acts as a potent oxytocin blocker, diminishing our ability to form deep, authentic connections. This chapter explores how modern life inhibits oxytocin production, the effects of this disruption on human relationships and well-being, and offers strategies for reclaiming genuine connection in a disconnected world.

How Modern Life Acts as an Oxytocin Blocker

The rapid advancement of technology and the increasing prevalence of digital communication have significantly altered the way we interact with one another. While these advancements bring convenience and access to information, they also contribute to an environment where oxytocin—the hormone responsible for connection, trust, and empathy—can be blocked. Below are the key factors in modern society that hinder oxytocin production:

1. Technology and Digital Disconnection

While technology connects us in unprecedented ways, it can also isolate us emotionally. Social media platforms, while facilitating communication, often create shallow interactions that fail to trigger the deep, trust-based connections necessary for oxytocin release. The superficial nature of online communication, characterized by likes, comments, and emojis, lacks the physical touch, eye contact, and genuine emotional exchanges that foster oxytocin production.

- **Impact**: Digital disconnection leads to social isolation. Constant engagement with screens, as opposed to face-to-face interactions, reduces opportunities for emotional bonding, leading to a deficiency in oxytocin and fostering feelings of loneliness and alienation.

- **Real-Life Example**: Emily, a college student, spent hours each day on social media, scrolling through posts and liking photos, but felt increasingly disconnected from her peers. Despite having thousands of "friends" online, she experienced a growing sense of loneliness. This is a direct result of the lack of meaningful, real-world interactions that stimulate oxytocin.

2. The Pressure of a 24/7, Always-On Culture

The rise of the "always-on" culture, enabled by smartphones and constant connectivity, has led to heightened stress and a reduction in the quality of personal interactions. The need to be perpetually available for work, social engagements, and online conversations contributes to emotional exhaustion, eroding the capacity for empathy and compassion, both of which are supported by healthy oxytocin levels.

- **Impact**: Chronic stress suppresses oxytocin production by increasing cortisol levels. The constant pressure to be responsive and productive leaves little room for genuine connection or emotional downtime, both of which are essential for oxytocin release.

- **Real-Life Example**: David, a corporate executive, felt constantly pressured to be available for calls and emails. His personal relationships suffered as he had less time to connect meaningfully with his family and friends. This constant state of alertness blocked his ability to experience moments of calm connection, reducing his oxytocin levels and increasing his stress.

3. Social Media and Validation-Seeking

The validation-driven nature of social media platforms, where self-worth is often measured in likes, followers, and comments, can create a disconnect from authentic social interaction. Social media interactions often prioritize external validation over internal fulfillment, leading to a cycle of superficial connection rather than true emotional bonding.

- **Impact**: When oxytocin is triggered by genuine connection and empathy, the validation-seeking behavior on social media can distort these natural processes. The result is an ongoing cycle of seeking external approval instead of fostering meaningful relationships that contribute to long-term emotional well-being.

- **Real-Life Example**: Sarah, an influencer, found herself obsessing over the number of likes and comments her posts received. Despite having a significant online following, she felt a deep sense of emptiness. Her reliance on external validation blocked her ability to connect emotionally with others, preventing oxytocin from being released during her interactions.

4. Urbanization and Physical Isolation

The increasing trend toward urban living, where people often live in high-rise apartments or densely populated areas, can contribute to a sense of isolation. Although cities offer a rich array of social and professional opportunities, urban environments can foster loneliness due to the lack of personal, intimate spaces for emotional connection. The sheer density of people, coupled with a focus on efficiency and productivity, leaves little room for the natural, face-to-face interactions that stimulate oxytocin.

- **Impact**: The absence of supportive, close-knit community structures can make individuals feel isolated even when surrounded by people. The lack of these vital connections inhibits the release of oxytocin, leading to social withdrawal and emotional numbness.

- **Real-Life Example**: Mike moved to a large city to pursue his career but found himself feeling increasingly disconnected. He worked long hours, rarely socialized, and didn't form close friendships. Despite being surrounded by thousands of people, he felt profoundly lonely—an outcome of the lack of meaningful, oxytocin-boosting interactions in his urban environment.

The Effects of Oxytocin Blockers in Modern Society

The modern environment, with its focus on speed, efficiency, and virtual connection, has far-reaching effects on human behavior and emotional well-being. When oxytocin is blocked, individuals may experience several negative consequences:

1. Decreased Empathy and Emotional Intelligence

Oxytocin is essential for empathy—the ability to understand and share the feelings of others. When oxytocin levels are blocked, individuals may find it harder to connect with others emotionally, leading to a decline in emotional intelligence. This can have serious consequences for both personal relationships and professional interactions.

Impact

2. Increased Loneliness and Social Isolation

A lack of meaningful human connection can lead to profound feelings of loneliness. Without oxytocin, which promotes bonding and social trust, individuals may become more withdrawn and isolated. This isolation can further block oxytocin release, creating a vicious cycle.

Impact

3. Decline in Relationship Quality

Oxytocin is critical for forming and maintaining close, trusting relationships, whether with family, friends, or romantic partners. When oxytocin blockers are present, these relationships can suffer. Couples may experience a breakdown in communication, trust, and intimacy, while family members may feel emotionally distant from one another.

Impact

4. Impaired Physical Health

Oxytocin is not only vital for emotional well-being but also for physical health. It plays a role in reducing inflammation, lowering blood pressure, and enhancing immune function. When oxytocin is blocked, the body can experience heightened levels of stress, leading to a range of physical health problems, including cardiovascular issues and a weakened immune response.

Impact

Reclaiming Connection in a Disconnected World

While modern society certainly presents significant challenges to oxytocin production, there are ways to counteract these blockers and restore connection, trust, and emotional well-being.

1. Fostering Face-to-Face Interactions

One of the most effective ways to combat digital disconnection is by prioritizing face-to-face communication. Spending time with friends, family, or colleagues in person can naturally boost oxytocin levels and promote deeper, more meaningful relationships.

Actionable Tip

2. Disconnecting from Technology

Taking regular breaks from screens, especially social media, can help reduce the negative effects of digital disconnection. Scheduling "digital detoxes" allows individuals to reset and refocus on in-person connections that are rich in oxytocin-triggering opportunities.

Actionable Tip

3. Creating Supportive Environments

Both at home and in the workplace, creating environments that encourage social bonding and trust can help counteract the effects of modern stressors. This can include regular team-building exercises, encouraging open communication, and prioritizing emotional well-being over constant productivity.

Actionable Tip

4. Practicing Empathy and Compassion

Cultivating empathy and compassion in daily life—whether at work, in social settings, or at home—helps stimulate oxytocin. Being present for others, listening actively, and offering support all create an atmosphere where oxytocin can flourish.

Actionable Tip

Conclusion

Modern society presents numerous challenges to oxytocin production, from digital disconnection to the pressures of urban life. However, by recognizing the impact of these blockers and actively taking steps to promote meaningful, face-to-face connections

Chapter 9: Oxytocin Blockers and Addiction

Addiction is a complex phenomenon that involves alterations in the brain's reward system. Central to these changes are the neurochemicals that regulate feelings of pleasure, connection, and well-being. Oxytocin, often referred to as the "bonding hormone," plays a critical role in establishing and reinforcing healthy connections, both social and emotional. However, when oxytocin levels are disrupted—either through blockers or by the impact of addictive behaviors—individuals may struggle with deeper connections, increasing the likelihood of addictive tendencies. This chapter explores how oxytocin blockers contribute to addiction, the specific mechanisms behind these interactions, and how understanding this relationship can lead to more effective treatment strategies.

The Relationship Between Oxytocin and Addiction

Addiction is often understood in terms of the brain's reward system, specifically the role of dopamine. When a person engages in addictive behaviors, such as substance abuse, compulsive gambling, or even certain eating habits, the brain releases dopamine as a reward. This neurotransmitter gives a feeling of pleasure, reinforcing the behavior. However, oxytocin also plays a vital, though less acknowledged, role in how we experience pleasure and connection. It is deeply linked to bonding, empathy, and trust—the very qualities that form the foundation of healthy relationships.

- **Impact of Blocked Oxytocin**: When oxytocin levels are low or blocked, individuals may experience a diminished ability to form deep emotional bonds or feel secure in their relationships. This emotional void may lead them to seek external sources of pleasure or relief, such as substances or compulsive behaviors, as a way to fill the gap. Essentially, the body turns to artificial stimuli to compensate for the lack of natural connection and satisfaction that oxytocin provides.

- **Real-Life Example**: John, a man in his early 30s, struggled with alcohol addiction for years. Despite numerous attempts at sobriety, he always relapsed. After a series of psychological evaluations, it was discovered that John had significant emotional trauma in his past, leading to chronic oxytocin depletion. His inability to form meaningful relationships left him emotionally isolated, making alcohol a temporary escape from the feelings of emptiness.

How Oxytocin Blockers Influence the Reward System

The reward system is not just about pleasure; it's also about motivation and the ability to experience meaningful connections. Oxytocin directly interacts with this system by helping to reinforce behaviors that strengthen emotional bonds. When oxytocin is blocked—whether by environmental stressors, trauma, or synthetic blockers—it can distort the reward system, pushing the individual to seek out unhealthy forms of "reward."

- **Impaired Reward Perception**: When oxytocin's positive effects are diminished, individuals may not derive pleasure from positive, social interactions. Instead, they may become more reliant on addictive behaviors or substances, which temporarily boost dopamine levels. This creates a pattern of behavior where the person continues seeking out external "rewards" to fill the emotional void left by a lack of oxytocin-driven bonding.

- **Real-Life Example**: Maria, a woman in her mid-40s, turned to overeating as a way to cope with chronic stress from her job. Despite having a supportive family, she found it difficult to connect emotionally with others, which in turn fueled her addiction to food. Her struggle was rooted in a deeper oxytocin blockage, which prevented her from fully engaging in nurturing, connection-based activities that would have triggered her natural reward system in a healthy way.

Addiction and Oxytocin Deficiency: A Vicious Cycle

The relationship between oxytocin and addiction is cyclical. On the one hand, addiction exacerbates the depletion of oxytocin—substance use or compulsive behaviors tend to reduce the brain's ability to produce this bonding hormone. On the other hand, a lack of oxytocin increases the likelihood of engaging in addictive behaviors, as individuals search for external substitutes for the deep connections and emotional bonding that are essential for healthy mental and physical well-being.

- **The Feedback Loop**: Chronic substance use or compulsive behavior can reduce the body's natural production of oxytocin, which in turn makes it harder for the individual to engage in healthy relationships. This creates a feedback loop in which the person becomes more isolated and dependent on the addictive behavior as a means of coping with the lack of connection. Over time, this cycle deepens, making addiction harder to break.

- **Real-Life Example**: Tom, a recovering opioid addict, struggled with the emotional aftermath of his addiction even after completing a rehab program. Despite his recovery from physical dependence, he still found himself feeling disconnected from his family and friends. He sought counseling, where it was revealed that his addiction had interfered with his oxytocin production, making it difficult for him to trust others or feel secure in relationships.

Addressing Addiction Through the Lens of Oxytocin

Understanding the role of oxytocin in addiction provides a new perspective on treatment and recovery. By addressing oxytocin blockers, we can better equip individuals to break free from the cycle of addiction and build healthier connections. Below are some strategies that can help restore oxytocin balance and facilitate recovery:

1. Therapeutic Bonding and Emotional Support

One of the most effective ways to increase oxytocin levels is through the establishment of secure, trusting relationships. Psychotherapy that focuses on emotional bonding—such as attachment-based therapy or trauma-informed care—can help individuals rebuild their capacity to trust and connect with others.

Actionable Tip

2. Oxytocin-Stimulating Activities

Certain activities have been shown to naturally increase oxytocin levels. These include physical touch (like hugging or massage), social bonding, and participation in activities that promote feelings of safety and connectedness. In the context of addiction recovery, engaging in these activities can help individuals feel more grounded and less reliant on addictive behaviors.

Actionable Tip

3. Physical Exercise and Social Connection

Exercise is a powerful tool in boosting oxytocin levels, particularly when combined with social interaction. Group activities such as fitness classes or outdoor adventures can foster a sense of camaraderie and trust, further reinforcing the brain's natural bonding mechanisms.

Actionable Tip

4. Mindfulness and Stress Reduction

Stress is one of the primary natural blockers of oxytocin. Mindfulness practices, which reduce stress and promote relaxation, have been shown to increase oxytocin production. By incorporating stress-reduction techniques into an addiction recovery plan, individuals can help alleviate the mental and emotional pressures that trigger addictive behaviors.

Actionable Tip

5. Nutritional Support

There is emerging evidence that certain foods can help boost oxytocin production, and this can be an important part of a recovery plan. Omega-3 fatty acids, found in foods like fish and flaxseeds, as well as the amino acid tryptophan (found in turkey, nuts, and seeds), are essential for maintaining healthy oxytocin levels.

Actionable Tip

Conclusion

Addiction is deeply entwined with the neurochemical balance in the brain, particularly when it comes to the role of oxytocin. By recognizing the impact that oxytocin blockers have on addictive behaviors, we can address addiction from a more holistic perspective —one that integrates the importance of connection, emotional bonding, and trust. Through therapeutic interventions, lifestyle changes, and social support, it is possible to overcome the negative effects of oxytocin blockers and break the cycle of addiction.

Chapter 10: The Science of Bonding and Attachment

Oxytocin is often called the "love hormone" because of its central role in facilitating bonding and attachment. It is released in response to social interactions, emotional experiences, and physical touch, and it fosters feelings of trust, affection, and connection. Attachment theory, developed by psychologist John Bowlby, posits that the bonds formed between individuals—particularly between parents and children—are crucial to emotional development and well-being. This chapter delves into the science of bonding and attachment, exploring how oxytocin plays a pivotal role in forming and maintaining these bonds, and how oxytocin blockers can disrupt this essential process.

The Role of Oxytocin in Attachment Theory

Attachment theory emphasizes the importance of secure, consistent, and nurturing relationships in early childhood, which serve as the foundation for healthy emotional development and interpersonal relationships throughout life. Central to this process is oxytocin, which acts as a chemical facilitator for bonding. It is particularly critical in the parent-child relationship but also plays a major role in romantic, familial, and social connections.

- **The Attachment System**: The attachment system refers to the behavioral and emotional patterns that govern how we form and maintain close relationships. In infants, the attachment system is activated when they are in need of care or comfort, prompting them to seek proximity to their caregivers. Oxytocin helps to reinforce these attachment behaviors by promoting feelings of safety and trust when a caregiver responds appropriately. This "secure base" allows children to explore the world and develop autonomy while knowing they can rely on their caregiver when needed.

- **Oxytocin and Secure Attachment**: Secure attachment is characterized by a sense of safety and comfort in the presence of a caregiver. When a child feels secure, oxytocin levels increase, creating a positive feedback loop. The caregiver's nurturing response to the child's needs releases oxytocin, which strengthens the bond between them, reinforcing trust and security.

- **Social Bonding and Oxytocin**: As individuals mature, oxytocin continues to play a key role in their ability to form strong social bonds. In romantic relationships, friendships, and even in the workplace, oxytocin helps foster empathy, mutual trust, and cooperation, all of which are essential for healthy, thriving relationships.

How Oxytocin Blockers Disrupt Bonds and Relationships

Oxytocin blockers—whether they are the result of trauma, chronic stress, negative environmental influences, or synthetic interventions—can interfere with the attachment process and disrupt the formation of strong, healthy bonds. When oxytocin is blocked, individuals may experience difficulty trusting others, forming meaningful emotional connections, and maintaining close relationships.

- **Trauma and Attachment Disruption**: Trauma, particularly early childhood trauma, can significantly disrupt oxytocin production. Children who experience neglect, abuse, or inconsistent caregiving may develop insecure or disorganized attachment styles. As a result, they may struggle to form healthy relationships later in life, often seeking out dysfunctional patterns of connection or avoiding closeness altogether. Trauma-related oxytocin blockers can lead to emotional isolation and difficulties in forming secure relationships as the brain's ability to trust and bond is compromised.

- **Chronic Stress and Relationship Strain**: Chronic stress also acts as an oxytocin blocker. In stressful situations, the body releases cortisol, the stress hormone, which inhibits the production and effectiveness of oxytocin. This can lead to emotional numbing, increased irritability, and difficulty connecting with others. Over time, the absence of oxytocin-driven bonding can erode relationships, causing feelings of loneliness and disconnection.

- **Synthetic Blockers and Social Impairment**: The use of certain medications, such as selective serotonin reuptake inhibitors (SSRIs) or other pharmaceutical treatments, may also interfere with the natural flow of oxytocin in the brain. While these drugs can be effective for managing mood disorders, they can sometimes dampen the emotional response to social interactions, impairing the ability to form deep, empathetic connections. This is particularly concerning in the context of romantic relationships, where the lack of oxytocin can lead to emotional detachment and intimacy issues.

- **Oxytocin Blockers and Emotional Intimacy**: Emotional intimacy is one of the key components of healthy relationships, and oxytocin is essential for nurturing this intimacy. When oxytocin levels are blocked or insufficient, individuals may struggle to feel emotionally close to others. In romantic relationships, this can lead to a sense of distance or disconnect, even if both partners are physically present. The inability to experience the emotional rewards of bonding can create a vicious cycle, leading to further emotional withdrawal and, in some cases, relationship breakdown.

Therapeutic Approaches to Repairing Disrupted Bonds

Fortunately, there are effective therapeutic strategies for repairing disrupted bonds and helping individuals overcome the effects of oxytocin blockers. Whether the disruptions stem from trauma, stress, or other sources, the following approaches can help restore healthy attachment patterns and strengthen emotional connections.

1. Attachment-Based Therapy

Attachment-based therapies, such as Emotionally Focused Therapy (EFT) and Dyadic Developmental Psychotherapy (DDP), are designed to help individuals and couples reconnect by addressing underlying attachment issues. These therapies focus on identifying and processing emotional wounds caused by past relational trauma and building secure, trusting connections.

Actionable Tip

2. Trauma-Informed Care

For individuals who have experienced significant trauma, a trauma-informed approach to therapy can help them process past wounds while rebuilding their capacity for connection. Trauma-informed care emphasizes safety, trust, and empowerment, which are critical for restoring healthy oxytocin levels. This approach can be especially useful for individuals with a history of insecure attachment or disrupted bonding.

Actionable Tip

3. Social and Emotional Support

One of the most powerful ways to increase oxytocin and foster attachment is through social and emotional support. Engaging in relationships that are nurturing, safe, and emotionally responsive can provide the stimulation needed to release oxytocin. This includes close friendships, family connections, and romantic relationships where mutual trust and care are established.

Actionable Tip

4. Mindfulness and Compassionate Practices

Mindfulness and compassion-based practices, such as loving-kindness meditation and mindfulness-based stress reduction (MBSR), can enhance the body's ability to produce oxytocin by reducing stress and fostering a sense of connection with oneself and others. These practices encourage an open-hearted approach to life, which facilitates the release of oxytocin.

Actionable Tip

5. Physical Touch and Nonverbal Communication

Physical touch is one of the most direct ways to stimulate oxytocin production. Simple actions like holding hands, hugging, or even casual touch during conversation can significantly increase oxytocin levels and improve bonding. Nonverbal communication, such as eye contact and attentive listening, also plays a crucial role in strengthening emotional bonds.

Actionable Tip

Conclusion

Oxytocin plays a pivotal role in the bonding and attachment processes that are central to human relationships. From infancy through adulthood, our capacity to form healthy emotional connections is deeply intertwined with the presence of this hormone. However, when oxytocin levels are blocked—due to trauma, stress, or other factors—attachment bonds can become disrupted, leading to emotional isolation and relationship difficulties. By understanding the science of bonding and attachment, we can adopt therapeutic approaches that help restore these connections and rebuild the natural flow of oxytocin, fostering healthier and more fulfilling relationships in all areas of life.

Chapter 11: Parenting and Oxytocin Blockers

Parenting is one of the most profound and transformative experiences in human life, shaping not only the child but also the parents themselves. At the heart of effective parenting lies a strong, secure bond between the parent and child. Oxytocin, often called the "bonding hormone," is a crucial component in this process. However, oxytocin blockers—whether due to stress, trauma, or other factors—can significantly impair the bonding process, impacting the parent-child relationship and the child's emotional development. This chapter explores the science of bonding during pregnancy and infancy, the impact of oxytocin blockers on parental attachment, and practical strategies for fostering secure attachment despite the presence of blockers.

The Science of Bonding During Pregnancy and Infancy

Bonding between parent and child begins even before birth, with oxytocin playing a central role in the initial connection. During pregnancy, the release of oxytocin in the mother's body is essential for labor and delivery, helping to facilitate contractions and the emotional connection to the baby. After birth, oxytocin continues to support the bond by promoting maternal behaviors like nursing, skin-to-skin contact, and responding to the baby's cues. This early bonding experience is foundational for the child's emotional and psychological development.

- **Prenatal Bonding**: Prenatal bonding is influenced by oxytocin, which encourages the mother to focus on the well-being of her unborn child. This connection is bolstered by physical sensations such as fetal movements, ultrasounds, and the awareness of the baby's development. Studies suggest that when a mother feels emotionally connected to her unborn child, oxytocin levels are higher, leading to more positive maternal behaviors post-birth.

- **Postnatal Bonding**: The birth process itself triggers a significant release of oxytocin, both in the mother and the infant. For mothers, this hormone facilitates emotional attachment and a deep sense of love for their newborns. For babies, early bonding experiences, such as breastfeeding and physical touch, stimulate the release of oxytocin, helping to establish a sense of safety and trust in the world around them.

- **The Role of Oxytocin in Parental Behavior**: Oxytocin not only fosters a connection between mother and baby but also enhances the father's role in parenting. Fathers, too, experience an increase in oxytocin levels when they interact with their newborns, especially through touch, eye contact, and shared caregiving tasks. This early interaction helps establish a foundation of trust, security, and emotional connection that supports the child's development and strengthens the parental bond.

The Impact of Oxytocin Blockers on Parental Attachment

While oxytocin is crucial for bonding, a variety of factors can act as blockers, impeding the release or effectiveness of this vital hormone. These blockers—whether stemming from stress, trauma, or external circumstances—can disrupt the natural attachment process between parents and children.

- **Stress and Anxiety**: Chronic stress, particularly during pregnancy or the early months of parenting, can interfere with oxytocin release. High levels of cortisol, the stress hormone, can inhibit oxytocin production, making it more difficult for parents to connect emotionally with their infants. Parents under stress may find it harder to respond to their child's needs with empathy and care, potentially leading to an insecure attachment style in the child.

- **Postpartum Depression (PPD)**: Postpartum depression is a common condition that affects many new mothers, and it can significantly block oxytocin production. Symptoms of PPD, such as feelings of hopelessness, fatigue, and emotional numbness, are often accompanied by a reduced ability to bond with the baby. This creates a cycle of emotional withdrawal, where the mother's inability to connect with the child further exacerbates feelings of isolation and sadness.

- **Trauma and Insecure Attachment**: Parents who have experienced childhood trauma, emotional neglect, or previous attachment wounds may find it difficult to form secure bonds with their children. These parents may be unconsciously blocked from feeling the joy and emotional satisfaction that comes with nurturing and bonding. Trauma can trigger a heightened state of vigilance or emotional detachment, making it harder to engage in caregiving behaviors that promote attachment.

- **Socioeconomic Stressors**: Financial instability, lack of social support, and challenging living conditions can also act as blockers to oxytocin. When parents are preoccupied with survival, they may struggle to give their full attention and emotional presence to their children. This, in turn, can lead to difficulties in forming strong, secure attachments.

- **Substance Abuse and Addiction**: Substance abuse can severely disrupt oxytocin levels, making it difficult for parents to bond with their children. Drugs or alcohol can dull emotional responses, reduce empathy, and impair the capacity for close, affectionate interactions. This can create an environment where oxytocin release is suppressed, leading to attachment disruptions in both parents and children.

Strategies for Fostering Secure Attachment Despite Blockers

Even when oxytocin blockers are present, there are ways to nurture the parent-child bond and promote secure attachment. The following strategies can help parents overcome the effects of blockers and create a strong, healthy relationship with their children:

1. Mindfulness and Stress Reduction Techniques

Practicing mindfulness and stress-reduction techniques, such as deep breathing, yoga, or meditation, can help reduce cortisol levels and encourage the natural release of oxytocin. Mindfulness practices help parents stay emotionally present and responsive to their child's needs, fostering a calm, attuned environment conducive to bonding.

Actionable Tip

2. Physical Touch and Skin-to-Skin Contact

Physical touch is one of the most powerful ways to stimulate oxytocin release. Skin-to-skin contact between parents and infants, whether through breastfeeding, holding, or gentle caressing, can greatly enhance bonding and attachment. For fathers or non-birthing parents, engaging in activities like carrying the baby in a sling or spending time with the infant on the chest can stimulate oxytocin and promote closeness.

Actionable Tip

3. Responsive Parenting

Being responsive to a child's emotional and physical needs fosters a sense of safety and trust. Responsive parenting involves paying attention to the child's cues, such as crying or seeking comfort, and addressing them with warmth and consistency. By meeting these needs, parents help regulate the child's stress response and promote the release of oxytocin.

Actionable Tip

4. Therapy and Counseling

Parents who struggle with past trauma, mental health issues, or postpartum depression can benefit from therapy or counseling. Therapists trained in attachment-based therapies can help parents understand the underlying emotional blocks and provide tools to rebuild their emotional connection with their children. Couples therapy may also be helpful if relationship stressors are contributing to attachment difficulties.

Actionable Tip

5. Social Support

Having a strong social support system is essential for promoting emotional well-being and fostering healthy attachment. Engaging with family, friends, or parenting groups can provide emotional encouragement and reduce isolation. Supportive relationships also provide opportunities for oxytocin release through social interactions, helping to balance the effects of stress.

Actionable Tip

6. Positive Parenting Practices

Fostering a positive, nurturing environment is crucial for oxytocin release. Encouraging your child with positive reinforcement, providing a sense of safety, and creating routines that emphasize love and care all help to establish secure attachment patterns. This approach strengthens the parent-child bond and fosters a sense of emotional security.

Actionable Tip

Conclusion

Parenting is a deeply emotional journey, and oxytocin plays a central role in forming and maintaining strong parent-child bonds. However, various factors can act as blockers, interfering with this natural process. By understanding the impact of oxytocin blockers and adopting strategies to overcome them, parents can foster secure attachment, even in the face of challenges. Through mindfulness, physical touch, responsive parenting, and social support, it is possible to nurture healthy, loving relationships with children that will provide the foundation for emotional well-being and growth throughout life.

Chapter 13: Oxytocin Blockers and the Workplace

The workplace is an environment where collaboration, trust, and communication are essential to achieving success. As the demands and dynamics of modern work culture evolve, so too do the challenges of fostering a productive and harmonious workplace. One key, yet often overlooked, factor that influences workplace culture is oxytocin. Known as the "bonding hormone," oxytocin plays a central role in social interactions, teamwork, and emotional regulation. Unfortunately, a variety of factors in the workplace can act as blockers to oxytocin production, hindering collaboration, innovation, and overall well-being. This chapter explores how oxytocin blockers manifest in professional settings, how they influence social dynamics and leadership, and strategies for cultivating a supportive, connected, and high-performing work environment.

The Effects of Oxytocin Blockers in Professional Settings

In the workplace, oxytocin is crucial for fostering positive social connections, teamwork, and emotional intelligence. It enhances feelings of trust, empathy, and cooperation, and helps individuals navigate the challenges of interpersonal interactions and conflict. However, workplace stressors—whether they stem from organizational culture, leadership styles, or individual challenges—can block oxytocin production, leading to negative consequences for employees and teams alike.

1. Stress and Overwork

Chronic stress is a pervasive issue in modern workplaces, with many employees facing high expectations, long hours, and increasing workloads. When stress levels are consistently high, cortisol, the body's stress hormone, suppresses the production of oxytocin. This creates a cycle where individuals may feel more isolated, disengaged, or emotionally distant from colleagues, even when they are physically present. Without the positive influence of oxytocin, the quality of communication and collaboration deteriorates, leading to burnout and a decline in overall morale.

- **Impact on teamwork**: High stress reduces the ability to build trust within teams. Team members may struggle to connect on an emotional level, affecting collaboration and innovation.
- **Impact on employee well-being**: A lack of oxytocin can contribute to increased feelings of anxiety, depression, and disconnection from the organization.

2. Toxic Work Cultures

Toxic work cultures, characterized by competition, lack of support, and poor leadership, act as significant blockers to oxytocin. When employees feel undervalued, unsupported, or fearful of retaliation, their ability to connect with colleagues is severely hampered. In such environments, trust is eroded, and individuals are less likely to engage in collaborative behaviors or show empathy toward others. Over time, this leads to a negative feedback loop, where the absence of oxytocin further fuels stress and disconnection.

- **Impact on relationships**: Toxic workplaces foster an environment where empathy and emotional intelligence are stifled. Employees may feel reluctant to express vulnerability or ask for help, fearing judgment or reprisal.
- **Impact on leadership**: Leaders who exhibit aggressive or authoritarian behavior create a climate of fear and distrust, blocking oxytocin and hindering their ability to connect meaningfully with their teams.

3. Lack of Social Interaction and Isolation

While the physical workplace environment has become more flexible and decentralized with the rise of remote work, the shift toward digital communication can create a sense of isolation. Remote workers, in particular, may miss out on the physical touchpoints—like informal conversations in the break room or shared moments of laughter—that trigger oxytocin release. Without these social interactions, employees may feel disconnected from their colleagues, leading to lower levels of engagement and a diminished sense of belonging within the organization.

- **Impact on team dynamics**: Virtual communication can sometimes hinder effective collaboration by limiting the spontaneous, emotional cues that come with face-to-face interactions. Video calls, for example, may lack the non-verbal cues necessary for building trust and rapport.

- **Impact on organizational culture**: When employees are physically isolated, they may feel disconnected from the organization's goals, values, and culture. This sense of separation can diminish motivation and affect job satisfaction.

How Social Dynamics and Leadership Are Influenced by Oxytocin Blockers

Effective leadership relies heavily on the ability to connect with others on a human level, inspire trust, and cultivate a shared sense of purpose. Leaders who are able to foster an environment where oxytocin can thrive encourage emotional intelligence, empathy, and collaborative problem-solving. Conversely, when leaders unknowingly block oxytocin production through toxic behaviors, lack of support, or ineffective communication, they undermine the very foundation of a productive workplace.

1. Leadership Styles and Oxytocin

Different leadership styles can either promote or inhibit oxytocin release. For example, transformational leaders, who emphasize inspiration, trust-building, and collaboration, tend to create environments where oxytocin flows freely. These leaders are skilled at fostering a sense of unity and belonging among team members, motivating them to contribute their best efforts to shared goals.

- **Transformational leadership**: This style encourages open communication, feedback, and recognition of individual contributions, all of which boost oxytocin levels and promote a sense of collective achievement.
- **Authoritarian leadership**: In contrast, a command-and-control leadership style, which focuses on micromanagement and top-down decision-making, can block oxytocin. Employees may feel mistrusted or undervalued, leading to disengagement and lower productivity.

2. Trust and Empathy in the Workplace

Trust is the cornerstone of effective collaboration, and oxytocin is the hormone that fosters this trust. Leaders and employees alike must feel confident that they can rely on one another, not just for completing tasks, but for emotional support as well. Oxytocin creates this bond of trust, allowing people to take risks, share ideas, and be vulnerable with one another.

- **Trust-building practices**: Transparency, consistency, and supportive behavior from leaders build trust and encourage open communication.
- **Empathy as a leadership skill**: Leaders who practice empathy—by actively listening to their employees, understanding their needs, and providing support—help foster an environment of trust and oxytocin release.

Strategies for Fostering a Supportive, Connected Work Environment

Given the importance of oxytocin in fostering social bonds, trust, and collaboration, organizations should prioritize strategies that encourage oxytocin production in the workplace. The following approaches can help break down the blockers to oxytocin and create an environment where employees feel connected, supported, and motivated to work together toward shared goals.

1. Create a Positive, Supportive Work Culture

A workplace culture that prioritizes emotional well-being, mutual respect, and collaboration promotes the release of oxytocin. Organizations can achieve this by implementing policies that encourage work-life balance, employee recognition, and open communication.

Actionable Tip

2. Encourage Social Interaction and Team–Building

While remote work can make social interactions more challenging, there are ways to cultivate a sense of community. Encourage team-building activities, both virtual and in-person, that allow employees to connect on a personal level. Casual conversations, team lunches, and group activities (like volunteering or wellness events) can stimulate oxytocin release and create a sense of belonging.

Actionable Tip

3. Provide Leadership Training on Emotional Intelligence

Leaders play a critical role in shaping the emotional environment of the workplace. By providing training on emotional intelligence (EI), organizations can equip their leaders with the tools to foster empathy, manage stress, and create a culture of trust.

Actionable Tip

4. Foster Open Communication

Clear and open communication is essential for building trust and reducing misunderstandings in the workplace. Encourage an environment where employees feel comfortable expressing their thoughts, ideas, and concerns without fear of judgment.

Actionable Tip

5. Prioritize Employee Well-Being

Reducing stress and promoting mental and physical well-being are critical for encouraging oxytocin production. Offering wellness programs, stress-management workshops, and resources for mental health can help reduce the blockers that hinder oxytocin release.

Actionable Tip

Conclusion

Oxytocin plays a pivotal role in creating a connected, collaborative, and high-performing workplace. However, when oxytocin blockers—such as stress, toxic culture, and isolation—take hold, the work environment suffers, and productivity, morale, and innovation can decline. By understanding the impact of oxytocin blockers and taking proactive steps to foster a supportive, trust-filled workplace, organizations can unlock the full potential of their teams. Prioritizing connection, empathy, and emotional intelligence is not only beneficial for individual well-being but is also a powerful strategy for driving organizational success.

Chapter 14: Oxytocin Blockers in Conflict Resolution

Conflict is an inherent part of human interaction. Whether in personal relationships, the workplace, or global diplomacy, disagreements and misunderstandings are inevitable. However, the ability to resolve conflict effectively and compassionately is essential for fostering deeper connections, trust, and collaboration. Central to the resolution process is the role of empathy, communication, and emotional regulation—skills that are profoundly influenced by oxytocin. As the "bonding hormone," oxytocin plays a key role in enhancing our capacity for empathy, promoting reconciliation, and diffusing tension. When oxytocin blockers are at play, however, conflict resolution becomes more challenging, as individuals may struggle to connect emotionally, leading to escalation rather than resolution. This chapter explores the impact of oxytocin blockers on conflict dynamics and offers strategies for reducing blockers and fostering understanding in tense situations.

How Oxytocin Blockers Contribute to Misunderstandings and Conflict

Oxytocin facilitates trust, empathy, and cooperative behavior—key elements needed for successful conflict resolution. When oxytocin levels are low or blocked, individuals are less likely to approach conflict with a mindset of collaboration and understanding. Instead, they may resort to defensive, aggressive, or protective behaviors that prevent meaningful dialogue.

1. Impaired Empathy and Emotional Regulation

Empathy allows us to understand and share the feelings of others, which is essential in conflict resolution. When oxytocin production is blocked, the brain's ability to recognize and resonate with the emotions of others diminishes. As a result, individuals are less likely to listen attentively or validate the emotions of others during a conflict. This lack of empathy can lead to misunderstandings, escalating emotions, and entrenched positions.

- **Impact on communication**: Without oxytocin's influence, people may interpret statements or actions more negatively, leading to increased tension. They may misread intent, assuming malice or disrespect where none exists.
- **Emotional distancing**: The absence of empathy makes it harder to regulate emotions. People may feel more reactive, defensive, or hostile, and less able to stay calm and constructive during disagreements.

2. Increased Susceptibility to Negative Bias

Oxytocin not only facilitates bonding but also helps counteract the tendency to see others as "other" or as adversaries. When oxytocin is blocked, people become more susceptible to negative biases, such as prejudice, stereotyping, and dehumanization. In the heat of conflict, this can manifest as a heightened "us vs. them" mentality, where parties are more focused on defending their own position than on understanding or empathizing with the other party's perspective.

- **Us vs. them mentality**: In conflicts, oxytocin's absence leads to greater polarization, where each side becomes entrenched in their perspective. This intensifies the adversarial nature of the conflict, making it more difficult to find common ground.

- **Failure to engage in perspective-taking**: Individuals with lower oxytocin may find it harder to see things from the other party's point of view, preventing a constructive resolution to the conflict.

3. Inability to Trust

Trust is the cornerstone of any productive conflict resolution. Oxytocin fosters the feeling of trust by increasing positive social bonds and encouraging open, honest communication. When oxytocin is blocked—whether through stress, fear, or negative social factors—trust becomes difficult to establish. Parties in conflict may feel defensive, suspicious, or guarded, hindering efforts to work together toward a resolution.

- **Impact on collaboration**: Trust is critical in collaborative problem-solving. Without it, individuals may be unwilling to compromise or work together to find mutually beneficial solutions.

- **Resistance to vulnerability**: Trust also allows individuals to be vulnerable, expressing their concerns and emotional needs openly. Oxytocin blockers make vulnerability feel risky, increasing the likelihood of defensive behaviors.

The Role of Oxytocin in Empathy and Reconciliation

Empathy is the emotional bridge that allows individuals to connect and resolve differences. It enables people to see conflict as an opportunity for understanding and growth, rather than as a threat. When oxytocin is present and functioning optimally, it enhances the brain's capacity for empathy and emotional attunement, key components of effective conflict resolution.

1. Oxytocin and the Power of Listening

Listening is one of the most powerful tools in conflict resolution. When individuals actively listen to one another, they create space for mutual understanding, fostering connection and reducing tension. Oxytocin is instrumental in this process, as it enhances the brain's ability to recognize the emotions and intentions behind words.

- **Active listening**: Oxytocin facilitates empathy-driven listening, allowing individuals to pick up on non-verbal cues (such as body language and tone of voice) and process the emotional significance of what is being said.
- **Validation**: Listening with empathy allows for emotional validation, which is essential for de-escalating conflict. When one party feels heard and understood, they are more likely to soften their stance and engage in cooperative dialogue.

2. The Role of Oxytocin in Emotional Reconciliation

Once empathy has been established, oxytocin fosters emotional reconciliation by promoting forgiveness, compassion, and a sense of shared humanity. Oxytocin's calming effects enable individuals to overcome anger or resentment, making it easier to forgive and restore relationships.

- **Forgiveness**: Oxytocin helps break the cycle of reactivity and defensiveness, creating space for forgiveness. When oxytocin levels are high, people are more likely to let go of grudges and seek reconciliation rather than holding onto anger.
- **Restoring connection**: Emotional reconciliation goes hand-in-hand with the restoration of emotional bonds. Oxytocin reinforces these bonds, ensuring that parties leave a conflict with a renewed sense of trust and closeness.

Tools for Reducing Oxytocin Blockers in Conflict Situations

Fortunately, there are several strategies that can help reduce oxytocin blockers during conflict resolution, allowing for more constructive, empathetic, and collaborative exchanges.

1. Breathing Techniques and Relaxation

Stress is one of the most significant blockers of oxytocin, as cortisol—the stress hormone—directly suppresses oxytocin production. During a conflict, both parties can engage in breathing techniques or relaxation exercises to reduce tension and foster a calmer, more receptive mindset.

- **Actionable Tip**: Try slow, deep breathing exercises during a conflict—inhale for four counts, hold for four, and exhale for four. This helps lower cortisol levels and increase the likelihood of oxytocin production.
- **Actionable Tip**: Guided relaxation or mindfulness exercises can also be beneficial, allowing both parties to regain emotional balance and approach the conversation with a clear mind.

2. Positive Language and Non-Verbal Cues

Language plays a powerful role in shaping the dynamics of conflict. Negative or accusatory language can trigger defensive responses, further blocking oxytocin. In contrast, positive language—such as using "I" statements instead of "you" statements—can reduce defensiveness and promote understanding. Non-verbal cues, such as maintaining eye contact or open body language, can also foster a sense of trust and connection.

- **Actionable Tip**: Use phrases like "I feel…" or "I need…" to communicate your own emotions and needs, rather than placing blame or making assumptions about the other person's intentions.
- **Actionable Tip**: Maintain open body language, avoid crossing arms, and make eye contact to convey openness and empathy.

3. Active Problem-Solving and Cooperation

In conflicts where both parties are motivated to find a solution, collaboration is key. Oxytocin facilitates cooperative behavior by strengthening social bonds and promoting a shared sense of purpose. Focus on finding common ground and working together toward a solution, rather than seeking to "win" the argument.

- **Actionable Tip**: Frame the conflict as a shared problem to be solved together, rather than a battle to be won. Encourage brainstorming solutions that benefit both parties.
- **Actionable Tip**: Take breaks if needed to give both parties time to cool down and reflect on possible solutions before resuming the discussion.

4. Physical Touch (When Appropriate)

Physical touch is a powerful way to release oxytocin and foster connection. A handshake, a pat on the back, or a gentle touch on the arm can signal trust and reduce tension during a conflict. However, physical touch should always be respectful and consensual, as it can be perceived differently depending on cultural norms and individual preferences.

Actionable Tip

Conclusion

Oxytocin plays a critical role in conflict resolution by fostering empathy, trust, and emotional connection. When oxytocin blockers—such as stress, defensiveness, or negative bias—are at play, conflict can escalate and become more challenging to navigate. By recognizing the impact of oxytocin blockers and employing strategies to reduce them, individuals can create an environment where conflict is seen as an opportunity for understanding, growth, and reconciliation. Whether in personal relationships, professional settings, or larger societal conflicts, mastering oxytocin blockers allows for more compassionate, effective conflict resolution that promotes deeper connections and stronger bonds.

Chapter 15: The Role of Oxytocin Blockers in Trauma Recovery

Trauma, whether psychological or physical, can have profound and lasting effects on an individual's mental and emotional health. Central to the recovery process is the ability to reconnect with others and foster a sense of safety and trust. Oxytocin, the hormone most often associated with bonding and trust, plays a pivotal role in this process, as it is deeply involved in emotional regulation, social bonding, and the body's healing response. However, trauma can block the production or effectiveness of oxytocin, creating a barrier to recovery. Understanding how oxytocin blockers manifest in trauma survivors and exploring strategies to heal and restore oxytocin balance is essential for overcoming the emotional and physiological impacts of trauma.

How Oxytocin Blockers Manifest in Trauma Survivors

Trauma, especially when experienced early in life or during critical periods of emotional development, can have a significant impact on the body's oxytocin system. Survivors of trauma often experience elevated stress levels, heightened anxiety, and difficulty in forming or maintaining healthy emotional connections. These experiences activate the body's natural defense mechanisms, such as the release of cortisol, which, when chronic, can suppress oxytocin production and block its positive effects.

1. Oxytocin's Role in Trauma and Healing

During times of stress or trauma, the body releases cortisol, which triggers a fight-or-flight response. While this is a natural survival mechanism, chronic stress can lead to prolonged elevations in cortisol, which, over time, negatively impacts oxytocin levels. The consequence is that the body's ability to experience empathy, trust, and social bonding becomes impaired. This creates a cycle where trauma survivors become emotionally distant, disconnected, and mistrustful, reinforcing their trauma.

- **Social isolation**: Trauma survivors may withdraw from relationships or have difficulty trusting others, both of which are linked to lower oxytocin levels. The emotional distance exacerbates feelings of isolation and loneliness, which can increase vulnerability to depression and anxiety.

- **Difficulty with emotional regulation**: Oxytocin helps regulate emotions and promote a sense of calm and safety. Without adequate oxytocin, trauma survivors may struggle with emotional outbursts, excessive worry, or emotional numbing, hindering their ability to heal and form healthy connections.

2. Impact of Trauma on Brain Function

Research has shown that trauma can alter the brain's neurochemistry, affecting areas of the brain responsible for emotional processing, memory, and social behavior. One key area of interest is the amygdala, which processes emotions and threat responses. Chronic trauma can cause the amygdala to become hyperactive, heightening the body's fear response and further suppressing oxytocin. Meanwhile, the prefrontal cortex, which regulates emotional control and social decision-making, may become less effective due to the long-term impacts of stress and trauma, further hindering the ability to regulate emotions or trust others.

- **Amygdala activation**: A heightened amygdala response causes increased fear and anxiety, triggering a cascade of stress responses that inhibit oxytocin production.

- **Reduced prefrontal cortex activity**: This leads to difficulty in exercising emotional regulation, leading to reactions that may appear irrational or disproportionate, which can make it harder for trauma survivors to interact positively with others.

The Healing Power of Oxytocin in Trauma Recovery

Despite the challenges that trauma presents, oxytocin also offers a potential pathway to healing. Oxytocin's role in social bonding, emotional regulation, and its ability to reduce stress positions it as a key player in trauma recovery. By reactivating the oxytocin system, trauma survivors can begin to rebuild trust, heal emotional wounds, and restore a sense of safety.

1. Oxytocin and the Restoration of Trust

One of the most important aspects of trauma recovery is the rebuilding of trust—both in oneself and in others. Oxytocin fosters a sense of security and attachment, which can help individuals feel safe enough to trust again. Therapeutic practices that enhance oxytocin production are crucial for overcoming the emotional walls that often develop after traumatic experiences.

- **Therapeutic relationships**: Building a trusting relationship with a therapist or counselor is often one of the first steps in recovery. This relationship can stimulate oxytocin release, creating a safe space for the survivor to process and heal their trauma.
- **Positive social interactions**: Engaging in supportive social networks, whether with friends, family, or peer support groups, can facilitate oxytocin release and encourage emotional healing.

2. Oxytocin and Emotional Regulation

Oxytocin's role in regulating emotions makes it a valuable tool for trauma survivors. Emotional regulation is essential for managing the overwhelming feelings that can arise during the recovery process, such as fear, shame, and anger. By increasing oxytocin levels, individuals can achieve a greater sense of emotional balance, allowing them to process painful memories and emotions without becoming overwhelmed.

- **Calming effects**: Oxytocin's ability to counteract the effects of stress hormones like cortisol helps trauma survivors calm their emotional responses, creating space for healthier coping mechanisms.
- **Improved resilience**: Higher levels of oxytocin are associated with increased resilience in the face of stress. Trauma survivors who can regulate their emotions more effectively are better equipped to navigate the challenges of recovery and rebuild their lives.

3. Physical Healing and Oxytocin

The impact of trauma isn't limited to emotional and psychological pain. Physical symptoms, such as chronic pain, tension, and fatigue, often accompany trauma. Oxytocin plays an important role in reducing inflammation, promoting healing, and enhancing the body's natural recovery processes. By boosting oxytocin, trauma survivors may experience a reduction in physical symptoms and an overall improvement in their well-being.

- **Reduction in pain**: Oxytocin has been shown to have analgesic effects, reducing the perception of pain. This can be especially beneficial for individuals dealing with chronic pain as a result of trauma.
- **Improved immune function**: Oxytocin's influence on the immune system helps the body recover from the long-term effects of trauma, supporting physical healing alongside emotional restoration.

Therapeutic Practices to Overcome Trauma-Related Blockers

While the impact of trauma on oxytocin levels can be significant, there are various therapeutic practices that can help restore the natural flow of oxytocin and facilitate recovery. These practices focus on rebuilding trust, improving emotional regulation, and reestablishing healthy social bonds.

1. Therapy and Trauma-Informed Care

Trauma-focused therapy approaches, such as Cognitive Behavioral Therapy (CBT), Eye Movement Desensitization and Reprocessing (EMDR), and somatic therapies, all aim to address the root causes of trauma while enhancing oxytocin production. These therapies create safe environments for survivors to process their experiences, express their emotions, and begin rebuilding trust in themselves and others.

- **Trauma-informed care**: It is crucial that mental health professionals use a trauma-informed approach that recognizes the unique needs of trauma survivors, creating a therapeutic space that fosters healing and supports the release of oxytocin.
- **Safe touch**: In certain therapeutic settings, safe and appropriate physical touch (such as gentle handholding or a reassuring touch) can activate oxytocin release, enhancing feelings of safety and trust.

2. Mindfulness and Meditation

Mindfulness and meditation practices have been shown to increase oxytocin levels by promoting relaxation and emotional awareness. These practices encourage trauma survivors to stay present in the moment, fostering self-compassion and reducing the impact of negative memories and emotions.

- **Mindfulness practices**: Techniques such as body scans, breathwork, and guided meditation can reduce stress and increase feelings of safety, providing trauma survivors with tools to regulate their emotions.
- **Loving-kindness meditation**: This specific type of meditation, which focuses on sending love and compassion to oneself and others, has been shown to increase oxytocin levels and improve emotional connection.

3. Social Support and Connection

Building a supportive social network is one of the most effective ways to combat oxytocin blockers in trauma recovery. Engaging in positive, nurturing relationships helps trauma survivors rebuild their social bonds and restore trust, both of which are critical for healing.

- **Support groups**: Peer support groups for trauma survivors offer a sense of belonging and validation, fostering oxytocin release through shared experiences and emotional connection.
- **Safe relationships**: Building new relationships or repairing existing ones with supportive, empathetic individuals can activate oxytocin and provide the emotional support necessary for healing.

Conclusion

Trauma recovery is a complex and multifaceted process, requiring not only emotional and psychological healing but also physical restoration. Oxytocin plays a central role in this healing journey, as it is integral to emotional regulation, trust-building, and social bonding. By understanding the impact of oxytocin blockers on trauma survivors and employing strategies to overcome them, individuals can begin to heal from their past wounds and reconnect with themselves and others in meaningful, positive ways. Through therapy, mindfulness, social support, and safe relationships, trauma survivors can restore oxytocin levels and unlock the healing power of connection, paving the way for a life of emotional well-being and resilience.

Chapter 16: Diet, Lifestyle, and Oxytocin Blockers

Our daily habits, including what we eat, how we live, and the choices we make about our environment, significantly impact our emotional and physical well-being. The production of oxytocin, often called the "love hormone," is no exception. From the food we consume to the way we handle stress, each element of our lifestyle has the potential to either promote or block the natural flow of oxytocin. In this chapter, we will explore the key factors that influence oxytocin levels and how diet and lifestyle choices can either support or hinder our ability to foster human connection, emotional regulation, and overall happiness.

The Connection Between Diet and Oxytocin

What we eat directly affects the balance of hormones in our body, and oxytocin is no exception. Certain nutrients and foods can enhance oxytocin production, while others may indirectly suppress it through their impact on the body's stress response or neurotransmitter balance.

1. Nutrients That Promote Oxytocin Production

Several vitamins, minerals, and amino acids play a key role in boosting oxytocin levels:

- **Magnesium**: Known for its calming and muscle-relaxing effects, magnesium also contributes to the regulation of oxytocin. Magnesium-rich foods like leafy greens, almonds, avocados, and bananas can help promote a sense of relaxation and increase oxytocin levels.

- **Vitamin C**: Vitamin C is critical in reducing stress and lowering cortisol, the hormone that can inhibit oxytocin. Foods rich in vitamin C, such as citrus fruits, strawberries, bell peppers, and kale, support the body's ability to produce oxytocin and reduce emotional stress.

- **Omega-3 Fatty Acids**: Omega-3s, found in fatty fish, flax seeds, and walnuts, are vital for brain health and neuroplasticity, supporting the production and function of oxytocin receptors in the brain.

- **Amino Acids**: Amino acids, particularly those found in protein-rich foods, such as tryptophan (found in turkey, nuts, and seeds), are precursors to serotonin, a neurotransmitter that influences oxytocin production. A diet rich in high-quality proteins ensures a steady supply of the building blocks necessary for oxytocin synthesis.

2. The Role of Hydration in Oxytocin Regulation

Hydration is often overlooked in discussions about hormone balance, but water plays an essential role in maintaining proper hormonal function. Dehydration can lead to stress and discomfort, triggering the release of cortisol and reducing oxytocin levels. By staying hydrated, you support your body's ability to regulate oxytocin and create the physical conditions for emotional well-being.

- **Water**: Water is the foundation of all physiological processes, including hormonal balance. Dehydration can lead to emotional instability, irritability, and stress, which can block oxytocin's positive effects.
- **Herbal teas**: Certain herbal teas, such as chamomile, peppermint, or lavender, can promote relaxation and help to reduce stress, indirectly supporting oxytocin production.

Lifestyle Habits That Affect Oxytocin Blockers

Beyond diet, several lifestyle habits influence oxytocin levels. From how we manage stress to the quality of our sleep, our daily routines can either help us foster positive connections or contribute to the blockers that hinder oxytocin production.

1. Stress Management

Chronic stress is one of the primary factors that inhibit oxytocin production. High levels of cortisol, the stress hormone, can block oxytocin receptors, making it difficult for the body to experience the positive effects of oxytocin, such as relaxation, trust, and connection.

- **Mindfulness**: Practicing mindfulness through techniques such as meditation, deep breathing, and focused awareness can significantly reduce stress and increase oxytocin production. Mindfulness helps activate the parasympathetic nervous system, promoting a state of calm and reducing the physiological effects of stress.

- **Relaxation techniques**: Activities like yoga, tai chi, and progressive muscle relaxation can also help lower cortisol levels and promote the flow of oxytocin by encouraging relaxation and emotional balance.

2. Quality Sleep

Sleep is a critical factor in both physical and emotional health. Insufficient sleep, particularly over extended periods, disrupts the production of key hormones, including oxytocin. Sleep deprivation can elevate cortisol levels, impair brain function, and reduce emotional regulation.

- **Sleep hygiene**: To support oxytocin production, it's important to practice good sleep hygiene. This includes creating a calm and dark environment, avoiding stimulating activities before bed, and ensuring adequate rest. Research shows that oxytocin levels rise naturally during deep, restorative sleep, particularly in the later stages of the sleep cycle.

- **Sleep and emotional processing**: During sleep, the brain consolidates memories and processes emotional experiences. Quality sleep supports the brain's ability to regulate emotions, making it easier to connect with others and repair any disruptions in social bonds.

3. Social Connection and Touch

Human connection is fundamental to oxytocin production. As a hormone that is deeply involved in social bonding, oxytocin thrives in environments of empathy, trust, and emotional closeness. The physical act of touch, including hugging, hand-holding, and cuddling, is one of the most effective ways to increase oxytocin levels.

- **Affectionate touch**: Simple acts of affection can trigger the release of oxytocin, promoting a sense of trust, connection, and emotional well-being. Physical closeness, whether in romantic relationships, friendships, or family bonds, fosters a safe environment that allows oxytocin to flourish.

- **Social bonding**: Engaging in positive social interactions, such as having meaningful conversations, participating in group activities, or even performing acts of kindness, can enhance oxytocin levels. Social engagement reduces stress and helps to regulate the emotional system, making it easier to form and maintain strong connections with others.

4. Time in Nature

Spending time outdoors in natural environments has been shown to reduce stress and improve overall well-being. Nature offers a sanctuary for relaxation, offering sensory experiences that calm the mind and promote emotional regulation.

- **Nature walks**: Walking in nature has been shown to reduce cortisol levels and elevate mood, which can foster the production of oxytocin. The simple act of being present in a natural environment helps reset the nervous system, encouraging a state of calm that allows for emotional connection and healing.

- **Exposure to sunlight**: Sunlight is a natural mood booster that helps regulate circadian rhythms and improve sleep. Sun exposure also promotes the production of serotonin, which plays a role in enhancing oxytocin function. This makes spending time outdoors in the daylight an important aspect of oxytocin regulation.

Creating a Lifestyle Plan to Increase Oxytocin and Decrease Blockers

To master oxytocin blockers and create a lifestyle that promotes emotional connection, it's essential to take a holistic approach that integrates diet, stress management, physical activity, sleep, and social connection. Here are some practical steps to create a balanced, oxytocin-boosting lifestyle:

1. Nutrient–Rich, Oxytocin–Boosting Diet

- Prioritize magnesium-rich foods, such as leafy greens, seeds, and almonds.
- Include omega-3 fatty acids in your diet through fatty fish, flaxseeds, and walnuts.
- Consume vitamin C-rich foods like oranges, strawberries, and bell peppers.
- Ensure adequate protein intake to support amino acid production, particularly tryptophan.

2. Stress Reduction and Mindfulness Practices

- Integrate mindfulness practices, such as meditation or yoga, into your daily routine.
- Set aside time for relaxation and self-care to combat daily stressors.
- Practice deep breathing or progressive muscle relaxation techniques to calm the nervous system.

3. Sleep Hygiene and Restorative Sleep

- Follow a consistent sleep schedule, ensuring 7-9 hours of restful sleep per night.
- Create a sleep-friendly environment by keeping your bedroom cool, dark, and quiet.
- Avoid screens and stimulating activities an hour before bed to prepare the body for sleep.

4. Regular Physical Activity

- Engage in regular exercise to improve mood, reduce stress, and stimulate oxytocin production.
- Incorporate a mix of aerobic, strength, and stretching exercises to optimize physical and emotional health.
- Try outdoor activities, such as hiking or biking, to combine exercise with the benefits of nature.

5. Social Engagement and Physical Touch

- Foster social connections by spending time with friends, family, and loved ones.

- Incorporate physical touch into your relationships, such as hugging, holding hands, or cuddling.

- Practice kindness and empathy to build trust and emotional intimacy in your relationships.

Conclusion

Diet, lifestyle, and environmental factors play a significant role in regulating oxytocin levels. By making mindful choices about what we eat, how we manage stress, and how we engage with others, we can create a lifestyle that fosters emotional well-being, connection, and resilience. The simple act of nurturing oxytocin through positive habits can enhance our ability to bond, empathize, and heal, creating a life that is more connected, compassionate, and fulfilling.

Chapter 17: Exercise and Oxytocin Blockers

Exercise is one of the most powerful tools for enhancing both physical health and emotional well-being. It not only strengthens the body but also plays a critical role in regulating the hormones that shape our social behavior, including oxytocin. For those struggling with oxytocin blockers, exercise can serve as an essential strategy to counteract the effects of these inhibitors. In this chapter, we will delve into the science behind how physical activity influences oxytocin levels, how it can mitigate the impact of blockers, and how you can design an exercise routine that fosters emotional resilience, connection, and overall health.

The Role of Physical Activity in Boosting Oxytocin Production

Exercise has been shown to stimulate the release of various neurochemicals, including endorphins, serotonin, and—most notably—oxytocin. The connection between physical activity and oxytocin is multifaceted, with several mechanisms at play:

1. **Stress Reduction**: Physical exercise acts as a potent stress reliever. Exercise reduces cortisol levels—the hormone associated with stress—that can otherwise block oxytocin's positive effects. By managing stress through physical exertion, the body creates a favorable environment for oxytocin to be released.

2. **Social Bonding Through Group Exercise**: Engaging in group exercise activities, such as team sports or group fitness classes, can amplify oxytocin release. Shared physical activity enhances feelings of social connection, trust, and belonging, further stimulating oxytocin production. The social nature of exercise fosters emotional bonds, reinforcing the positive effects on mental health and well-being.

3. **Physical Touch and Oxytocin**: Some forms of exercise, such as partner yoga, massage therapy, or martial arts, incorporate physical touch, which directly stimulates oxytocin production. The benefits of touch—whether in a therapeutic or social context—are well-documented, as they create a sense of security and comfort that enhances emotional well-being.

4. **Neuroplasticity and Emotional Resilience**: Regular physical activity supports brain health by promoting neuroplasticity, the brain's ability to adapt and reorganize itself. Neuroplasticity plays a key role in emotional resilience, allowing individuals to better cope with stress, trauma, and emotional difficulties. As oxytocin is essential for emotional regulation, its increase through exercise helps maintain a stable mood, reduces anxiety, and improves emotional bonding.

Exercise can serve as a direct antidote to the disruptive effects of oxytocin blockers, such as those caused by chronic stress, trauma, or social isolation. These blockers may suppress oxytocin production and hinder the body's ability to experience the positive effects of connection, empathy, and trust. Here's how exercise specifically addresses these challenges:

1. **Counteracting Chronic Stress**: Chronic stress is one of the primary culprits behind elevated cortisol levels, which can block oxytocin. Exercise helps break the cycle of stress by lowering cortisol and activating the parasympathetic nervous system, the body's "rest and digest" mode. By balancing cortisol levels, exercise makes it easier for oxytocin to flow freely.

2. **Repairing the Damage of Trauma**: Trauma, especially when unresolved, can cause oxytocin blockers to manifest as emotional numbness, anxiety, or social withdrawal. Exercise, particularly aerobic activities like running, swimming, or cycling, has been shown to improve mood and reduce symptoms of anxiety and depression, making it a vital tool in trauma recovery. Regular physical activity can help survivors of trauma reconnect with their bodies, emotions, and social support systems, ultimately aiding in the healing process.

3. **Combating Isolation**: Loneliness and social isolation are common factors that decrease oxytocin levels. By engaging in physical activities that involve others—whether in a community class, social sports leagues, or even group hikes—individuals can foster social connections and trigger oxytocin release. Even just spending time with a workout partner can enhance feelings of connection and belonging, directly countering the negative effects of isolation.

4. **Promoting Positive Emotional States**: Exercise has a unique ability to induce "flow states"—a heightened sense of focus and connection to the present moment. These states are not only emotionally rewarding but also physiologically beneficial, stimulating the release of oxytocin, serotonin, and other "feel-good" hormones. By regularly exercising, individuals can cultivate emotional balance and a positive outlook, even in the face of external challenges or internal blockers.

Designing an Exercise Routine to Balance Oxytocin Levels

To maximize the benefits of exercise in regulating oxytocin, it's important to choose activities that promote not only physical health but also social connection, relaxation, and emotional resilience. Below are strategies for designing an exercise routine that works to balance oxytocin levels and overcome blockers:

1. **Incorporate Cardiovascular Exercise**: Aerobic exercises such as running, cycling, swimming, and brisk walking are excellent for reducing stress, improving heart health, and boosting mood. Aim for at least 30 minutes of moderate-intensity cardio most days of the week. These exercises help regulate hormones and trigger the release of oxytocin, endorphins, and serotonin.

2. **Strength Training for Mental Clarity**: While aerobic exercise is crucial for mood regulation, strength training is equally important. Lifting weights or engaging in bodyweight exercises helps to boost self-esteem, foster resilience, and support overall mental health. Strength training promotes a feeling of accomplishment and empowerment, which can positively influence oxytocin levels.

3. **Group Exercise Activities**: Participating in group activities like yoga, dance classes, or team sports can amplify the benefits of exercise by incorporating social interaction. These activities not only improve physical fitness but also help you bond with others, increasing the release of oxytocin. Group workouts foster a sense of community, trust, and emotional connection.

4. **Mind-Body Practices**: Activities like yoga, tai chi, or Pilates are particularly effective at regulating both body and mind. These practices incorporate mindfulness and breathwork, helping to reduce stress and promote relaxation. As a result, these exercises can increase oxytocin levels while promoting physical flexibility, emotional calm, and mental clarity.

5. **Physical Touch Through Exercise**: Exercises that involve physical touch or connection—such as partner yoga, massage therapy, or even martial arts—can significantly enhance oxytocin levels. The tactile nature of these activities fosters trust and emotional bonding, which are essential for overcoming emotional blockers.

6. **Nature-Based Exercises**: Activities performed outdoors, such as hiking, cycling, or walking in a park, combine the benefits of exercise with the calming effects of nature. Natural settings reduce cortisol levels, enhance mood, and stimulate the production of oxytocin, providing a holistic approach to overcoming blockers.

7. **Consistency is Key**: The most important factor in using exercise to balance oxytocin is consistency. Regular physical activity—done consistently over time—produces cumulative benefits that help to regulate hormones, reduce stress, and improve emotional connection. Aim to establish a routine that includes a mix of cardiovascular, strength, and mind-body activities, with an emphasis on enjoyable, stress-relieving exercises.

Conclusion

Exercise is a powerful tool for boosting oxytocin production and counteracting the impact of oxytocin blockers. Whether through cardiovascular workouts, strength training, group activities, or mind-body practices, physical activity not only strengthens the body but also nurtures the emotional and social connections essential for well-being. By incorporating regular exercise into your routine, you can support the natural flow of oxytocin, reduce stress, and improve emotional resilience, paving the way for deeper connections and greater emotional health.

Chapter 18: Meditation, Mindfulness, and Oxytocin

In the fast-paced, high-stress world we live in, it's easy to become disconnected from ourselves and others. Modern life, with its constant demands and distractions, often leads to reduced oxytocin levels and increased emotional barriers. In this context, meditation and mindfulness practices can be powerful tools to help regulate oxytocin, combat the effects of blockers, and restore a deeper connection to ourselves and those around us. This chapter will explore how mindfulness and meditation practices influence oxytocin production, how they counteract blockers, and how to incorporate these practices into your daily routine for optimal emotional health.

The Connection Between Mindfulness Practices and Oxytocin Regulation

Mindfulness is the practice of being fully present and engaged in the moment, without judgment or distraction. It involves observing your thoughts, emotions, and bodily sensations with awareness, cultivating an open and non-reactive state of mind. Research has shown that mindfulness practices, including meditation, can directly influence the levels of oxytocin in the body and brain, thereby fostering emotional regulation, empathy, and connection.

1. **Stress Reduction and Oxytocin**: One of the primary benefits of mindfulness is its ability to reduce stress. When we're stressed, our bodies release cortisol, a hormone that can inhibit oxytocin production and block its positive effects. By practicing mindfulness, we activate the body's relaxation response, lowering cortisol and promoting the release of oxytocin. This shift in neurochemistry can help restore emotional balance and encourage feelings of connection and well-being.

2. **Enhancing Empathy and Compassion**: Mindfulness practices, particularly those focused on compassion and loving-kindness (such as "Metta" meditation), have been shown to increase oxytocin levels. These practices involve focusing on sending positive, loving thoughts and wishes to others, which can strengthen emotional bonds and deepen feelings of empathy. The more we practice empathy, the more oxytocin is released, creating a positive feedback loop that enhances both our emotional intelligence and social interactions.

3. **Regulation of Negative Emotions**: Mindfulness teaches us to observe our emotions without judgment, helping us to respond rather than react. This non-judgmental awareness is crucial for managing emotions such as anger, fear, or resentment, which can act as oxytocin blockers. By developing the ability to observe our emotions from a place of acceptance, we allow the body to process and release negative feelings without suppressing them. This promotes the free flow of oxytocin and supports emotional healing.

4. **Neuroplasticity and Oxytocin**: Just as physical exercise promotes neuroplasticity, mindfulness also enhances brain function and emotional resilience through neuroplasticity. Regular mindfulness practice has been shown to physically alter the brain, particularly in areas related to emotional regulation, empathy, and social connection. As the brain becomes more adept at regulating emotions, the effects of oxytocin blockers are mitigated, and we become more capable of fostering deeper, more meaningful relationships.

How Meditation Helps Combat Oxytocin Blockers

Meditation, a core component of mindfulness, is one of the most effective ways to counteract the effects of oxytocin blockers. Meditation helps lower stress, regulate emotions, and foster a sense of calm—all of which are essential for maintaining healthy oxytocin levels. Below are some key ways meditation specifically addresses the challenges posed by oxytocin blockers:

1. **Combatting Stress and Trauma**: Meditation helps individuals with chronic stress, anxiety, and trauma by creating a safe mental space for healing. Oxytocin blockers often stem from unresolved emotional wounds, such as past trauma or prolonged stress. Meditation, particularly guided trauma-sensitive or body-awareness practices, can help release stored tension in the body, reduce the impact of past trauma, and gradually restore oxytocin levels.

2. **Cultivating Self-Compassion**: One of the most transformative benefits of meditation is the cultivation of self-compassion. Oxytocin is crucial in the bonding process not only with others but also with oneself. Self-compassion meditation involves offering kindness and understanding to oneself, reducing self-criticism and shame, which are common emotional barriers that block oxytocin. This practice helps individuals develop a nurturing relationship with themselves, which enhances their ability to form compassionate bonds with others.

3. **Fostering Connection**: Meditation can help foster a sense of connection with others by increasing our ability to be present and fully engaged in our relationships. When we are distracted or overwhelmed, we may struggle to connect with others emotionally, which can lower oxytocin levels and create a cycle of disconnection. Meditation practices, such as loving-kindness or compassion meditation, encourage us to cultivate a positive mindset toward others, thereby increasing oxytocin and improving interpersonal relationships.

4. **Enhancing Emotional Resilience**: Meditation builds emotional resilience by teaching individuals to observe their thoughts and emotions without becoming overwhelmed by them. This ability to detach from emotional reactivity helps break the cycle of emotional stress and reactivity, which are key factors in the inhibition of oxytocin production. As individuals learn to regulate their emotional responses, they become more open to connection and less likely to experience emotional shutdowns that block oxytocin.

Techniques for Cultivating Oxytocin Through Mindfulness

There are many ways to incorporate mindfulness practices into your daily life that will foster oxytocin production. Here are several mindfulness techniques to cultivate oxytocin and combat blockers:

1. **Loving-Kindness Meditation (Metta)**: This practice involves silently repeating phrases of goodwill and loving-kindness towards yourself and others. By wishing happiness, health, and well-being to yourself, loved ones, and even neutral or difficult people, you increase the flow of oxytocin. The focus on kindness and positive intention strengthens bonds and reduces feelings of isolation, which are common causes of oxytocin blockers.

2. **Breath Awareness Meditation**: This simple yet powerful practice involves focusing your attention on your breath. As you inhale and exhale, pay close attention to the sensations in your body, allowing yourself to become present in the moment. This practice reduces stress, calms the nervous system, and enhances emotional balance, promoting a healthy environment for oxytocin to flow.

3. **Body Scan Meditation**: In this practice, you mentally scan your body from head to toe, paying attention to areas of tension or discomfort. This practice helps increase body awareness and release stored emotional tension, which can block oxytocin. The process of consciously relaxing each part of your body can enhance your overall sense of well-being, reduce stress, and promote feelings of connection.

4. **Gratitude Meditation**: This practice involves focusing on the positive aspects of your life, taking time to express gratitude for the people, experiences, and things that bring you joy. Gratitude meditation can foster a sense of abundance, help break down emotional barriers, and increase oxytocin production. When we focus on the positive, we open ourselves up to greater feelings of connection and affection toward others.

5. **Mindful Movement**: Practices like yoga, tai chi, or walking meditation combine mindfulness with gentle movement. These activities foster a connection between the body and mind, enhancing awareness and promoting oxytocin production. Regular practice of mindful movement can also help release tension, reduce stress, and improve mood.

Conclusion

Meditation and mindfulness practices are essential tools for restoring balance and regulating oxytocin levels. By reducing stress, enhancing emotional resilience, and cultivating compassion, these practices help counteract the effects of oxytocin blockers, making it easier to form healthy, supportive connections. Whether through breath awareness, loving-kindness meditation, or mindful movement, incorporating these techniques into your daily life can improve your emotional well-being, deepen your relationships, and create a life full of meaningful connections. With consistent practice, you can master the art of mindfulness and harness the power of oxytocin to foster a healthier, more connected existence.

Chapter 19: Oxytocin Blockers in Healthcare and Therapy

In the journey of healing and emotional well-being, the role of healthcare providers is indispensable. When it comes to overcoming the effects of oxytocin blockers, both physical and emotional health professionals must recognize the complex interplay of neurochemistry, behavior, and interpersonal dynamics. In this chapter, we will explore how healthcare providers can identify and address oxytocin blockers in their patients, the therapeutic approaches available to support patients in overcoming these blockers, and real-life case studies that demonstrate the success of such interventions.

The Role of Healthcare Providers in Addressing Oxytocin Blockers

Healthcare providers—ranging from physicians to psychologists to alternative medicine practitioners—are crucial players in the management of oxytocin blockers. They can identify the signs of oxytocin disruption and provide targeted interventions to mitigate the effects, ultimately restoring emotional health, improving relationships, and promoting social connectedness.

Diagnosis and Identification

Indicators of Oxytocin Blockers

Holistic Care Approach

- *Physical Care*: Providers can help manage underlying physical factors that affect oxytocin production, such as sleep, nutrition, and exercise. They can also identify and treat underlying conditions that might contribute to oxytocin blockers, such as chronic pain or illness.

- *Psychological Support*: In addition to physical care, therapists play a key role in supporting patients with oxytocin disruptions. Therapeutic interventions can help address the emotional and cognitive factors that may be contributing to the blockage. Cognitive Behavioral Therapy (CBT), trauma-focused therapy, and somatic therapies can all help patients re-regulate their emotional responses, creating a more open space for oxytocin production.

Collaborative Care

Therapeutic Approaches to Overcoming Oxytocin Blockers

Once oxytocin disruption has been identified, there are several therapeutic approaches that can be used to help restore balance and facilitate emotional connection. These treatments may involve a combination of traditional therapies, alternative approaches, and lifestyle modifications.

1. **Cognitive Behavioral Therapy (CBT)**: CBT is one of the most widely used therapeutic approaches for addressing the negative thought patterns and behaviors that contribute to emotional distress. Since oxytocin blockers often arise from distorted thinking—such as lack of trust, fear of vulnerability, or avoidance of connection—CBT can help individuals reframe these thoughts and develop healthier ways of interacting with others. By challenging negative beliefs and introducing more adaptive emotional responses, CBT can help reduce the emotional triggers that block oxytocin.

2. **Trauma-Informed Therapy**: For many individuals, oxytocin blockers stem from past trauma. In these cases, trauma-informed therapy is essential. This therapeutic approach recognizes the long-term impact of trauma on the brain and body and incorporates methods designed to help individuals safely process and heal from their past experiences. Techniques such as EMDR (Eye Movement Desensitization and Reprocessing), somatic experiencing, and other body-based therapies can be particularly effective in restoring emotional balance and fostering connection by releasing stored trauma that blocks oxytocin.

3. **Attachment-Based Therapy**: Since oxytocin plays a critical role in forming and maintaining healthy attachments, attachment-based therapy can be an important tool for individuals with disrupted bonds. This therapy focuses on re-establishing secure emotional connections, whether with a partner, parent, or child. By providing a safe environment for emotional expression and healing, attachment-based therapy helps individuals develop the trust and empathy necessary for strong, oxytocin-enhancing relationships.

4. **Mindfulness and Meditation**: Mindfulness practices, including meditation, are powerful tools for overcoming oxytocin blockers. As outlined in Chapter 18, mindfulness and meditation can increase oxytocin levels by reducing stress, cultivating compassion, and improving emotional regulation. Integrating mindfulness practices into therapy sessions or encouraging patients to engage in regular mindfulness exercises can significantly improve their emotional well-being and help foster deeper, more empathetic connections.

5. **Social and Emotional Skills Training**: For individuals who struggle with social connection due to oxytocin blockers, social and emotional skills training can be invaluable. These programs focus on teaching individuals how to navigate social interactions, build trust, and express vulnerability in healthy ways. Role-playing exercises, communication skills, and emotional awareness techniques can be employed to help patients overcome the fear or hesitation that prevents them from forming meaningful relationships.

6. **Group Therapy and Social Connection**: Sometimes, oxytocin blockers manifest in social isolation. In these cases, group therapy can provide a supportive environment where individuals can practice connecting with others in a safe and guided way. Group therapy offers the added benefit of shared experiences, which can help individuals feel less alone in their struggles. Group support fosters the release of oxytocin by facilitating emotional bonding with others who understand and empathize with their challenges.

7. **Pharmacological Interventions**: While not typically the first line of defense, in some cases, medications may be considered to help manage symptoms related to oxytocin blockers. For example, medications that target anxiety, depression, or trauma may help regulate emotional responses and create a more receptive environment for oxytocin production. However, this approach should be carefully monitored, and any pharmaceutical intervention should be paired with behavioral therapies for optimal results.

Case Studies: Successful Interventions

Case Study 1: Healing Through Attachment-Based Therapy

Sarah, a 35-year-old woman, had struggled with deep feelings of loneliness and disconnection for most of her life. Despite being surrounded by supportive friends and family, she often felt emotionally isolated, unable to form deep, trusting connections. After a series of failed romantic relationships, Sarah sought therapy. Through attachment-based therapy, she explored the roots of her disconnection, which stemmed from early childhood trauma and a lack of emotional support from her parents. Over the course of therapy, Sarah learned to develop healthier attachment patterns and gradually began to experience more trust and connection with others. Her relationships, both romantic and platonic, improved, and her levels of oxytocin began to balance out.

Case Study 2: Overcoming Trauma with Somatic Experiencing

John, a 40-year-old man, came to therapy after struggling with chronic anxiety, emotional numbness, and an inability to trust others. His symptoms had worsened after a traumatic event in his childhood, leaving him with deep emotional wounds that blocked his ability to connect with those around him. Somatic experiencing therapy helped John process the trauma stored in his body, releasing emotional tension and physical sensations tied to past trauma. As a result, John experienced a marked reduction in anxiety, and his ability to form connections improved. His oxytocin levels rose as he began to feel more relaxed, open, and connected with others.

Conclusion

Oxytocin blockers represent a significant challenge to emotional health and well-being, but with the right therapeutic interventions, they can be effectively addressed. Healthcare providers have a critical role to play in diagnosing and treating these blockers, utilizing a variety of therapeutic approaches tailored to the individual. From trauma-informed therapy to mindfulness practices, and from attachment-based therapy to social skills training, there are a wealth of strategies available to help individuals restore their oxytocin balance and re-establish deep, meaningful connections. By adopting a holistic, collaborative approach, healthcare providers can support their patients in overcoming oxytocin blockers and unlocking the power of human connection.

Chapter 20: Advancements in Oxytocin Blocker Research

In the ever-evolving field of neuroscience and human behavior, the exploration of oxytocin and its blockers remains one of the most dynamic areas of research. Understanding how these blockers influence human interaction, bonding, and mental health is essential for creating effective interventions and treatments. This chapter will explore the latest findings in oxytocin blocker research, highlight the breakthroughs that are shaping our understanding of the hormone, and address the ethical considerations surrounding the use of oxytocin in clinical settings.

Current Research on Oxytocin Blockers and New Findings

As research on oxytocin continues to unfold, new studies are shedding light on the intricate mechanisms that influence both the production of oxytocin and its blockers. Several key areas of focus are driving advancements in our understanding of these complex neurochemical processes.

Genetic and Epigenetic Factors

The Role of Stress

Oxytocin and the Brain's Reward System

Oxytocin as a Social Reward

Oxytocin Blockers in Mental Health Disorders

Oxytocin and PTSD

Pharmaceutical and Synthetic Oxytocin Blockers

Targeted Therapeutics

Breakthroughs in Treatment and Understanding

As our understanding of oxytocin blockers deepens, several breakthroughs have the potential to reshape how we treat mental health disorders, trauma, and social disconnection:

Oxytocin Receptor Agonists

Applications in Autism

Oxytocin-Boosting Interventions

Social Support Networks

Ethical Considerations in Advancing Oxytocin Blocker Research

As with all advancements in neuroscience, the research surrounding oxytocin blockers raises several ethical questions, particularly in the context of using oxytocin or its blockers as therapeutic agents.

1. **Manipulating Human Connection**

   The use of oxytocin to enhance human connection and emotional bonding may offer profound benefits, but it also raises concerns about the potential for manipulation. Could oxytocin agonists be used to artificially create feelings of trust and empathy in situations where they may not be appropriate, such as in marketing or coercive relationships? The ethical implications of using oxytocin to influence human behavior must be carefully considered, and ethical guidelines should be established to govern its use.

2. **Long-Term Effects of Synthetic Oxytocin**

   While synthetic oxytocin has proven effective in clinical settings (e.g., for inducing labor or managing PTSD), its long-term effects on the brain's oxytocin receptors are not yet fully understood. If used excessively or inappropriately, synthetic oxytocin could alter the brain's natural neurochemical balance, leading to unintended consequences. Long-term studies are needed to assess the safety of these treatments and ensure that they do not disrupt the natural processes that support emotional and social well-being.

3. **Access and Equity**

Finally, as oxytocin-based therapies move closer to clinical application, there are concerns about equitable access. These treatments could become costly or inaccessible to certain populations, exacerbating disparities in mental health care. Ensuring that all individuals, regardless of socioeconomic status, have access to these innovative therapies will be essential to their success in improving human connection and behavior.

Conclusion

Advancements in oxytocin blocker research are bringing us closer to understanding the fundamental mechanisms of human connection, bonding, and emotional regulation. From genetic studies to pharmaceutical breakthroughs, the research landscape is ripe with potential for improving mental health and fostering deeper, more meaningful relationships. However, as with all emerging fields, the ethical considerations surrounding these advancements must be handled with care to ensure that the power of oxytocin is used responsibly. As we continue to explore the science of oxytocin, we are unlocking new pathways to emotional well-being, deeper connection, and the healing power of human interaction.

Chapter 21: Building Emotional Intelligence in the Face of Oxytocin Blockers

Emotional intelligence (EI) is the ability to recognize, understand, manage, and influence our own emotions, as well as the emotions of others. It is a cornerstone of human connection, empathy, and effective communication. In the context of oxytocin blockers, emotional intelligence becomes both a challenge and a solution. This chapter explores how oxytocin blockers impact emotional intelligence and offers practical tools to enhance empathy, self-awareness, and emotional regulation, even in the face of these neurochemical obstacles.

How Oxytocin Blockers Affect Emotional Intelligence

Oxytocin is often referred to as the "love hormone" because of its critical role in promoting trust, social bonding, and empathy. It directly influences our ability to read others' emotional states and respond with appropriate emotional reactions. When oxytocin levels are blocked—either due to natural stress responses, trauma, or environmental factors—individuals may struggle with recognizing emotional cues, responding compassionately, and maintaining emotional regulation.

1. **Impaired Empathy**: Oxytocin is essential for empathy, allowing individuals to understand and share the feelings of others. Blockers disrupt this ability, leading to emotional detachment, difficulty in forming connections, and even a lack of concern for others' well-being. In relationships, this can manifest as insensitivity, miscommunication, and emotional isolation.

2. **Reduced Self-Awareness**: Emotional intelligence begins with self-awareness—recognizing and understanding one's own emotions. Oxytocin blockers, particularly in high-stress environments or due to trauma, can create a disconnect between our emotional experiences and our conscious recognition of those feelings. This can lead to difficulties in identifying personal needs, managing emotional responses, and maintaining emotional balance.

3. **Impaired Emotional Regulation**: Oxytocin plays a role in regulating emotional responses, helping us stay calm in stressful situations. When blocked, emotional regulation becomes challenging. This can lead to overreactions, frustration, anxiety, or even depression—all of which disrupt emotional intelligence and hinder healthy social interactions.

Training and Tools to Improve Empathy and Self-Awareness

Fortunately, emotional intelligence is not static. It can be developed and enhanced, even in the presence of oxytocin blockers. By actively engaging in strategies that counteract the effects of these blockers, individuals can strengthen their ability to connect with others and better manage their own emotional landscapes.

Mindful Reflection and Emotional Awareness

Practical Exercise

Empathy Training and Active Listening

Practical Exercise

Cognitive Behavioral Techniques (CBT) for Emotional Regulation

Practical Exercise

Building Emotional Agility

Practical Exercise

The Importance of Emotional Intelligence in Overcoming Oxytocin Blockers

Mastering emotional intelligence is not only a path to better emotional regulation and social interactions—it is also a key strategy for overcoming the effects of oxytocin blockers. The stronger our emotional intelligence, the more we are able to counteract the negative effects of stress, trauma, and environmental disconnection. Emotional intelligence helps us reclaim our capacity for connection, even in the face of obstacles.

1. **Improved Relationships**: High emotional intelligence leads to stronger, more empathetic relationships. It allows individuals to manage their own emotions and respond more thoughtfully to others, reducing the likelihood of misunderstandings or conflict. This can counteract the disconnection caused by oxytocin blockers and foster deeper, more meaningful bonds.

2. **Resilience in Challenging Situations**: Developing emotional intelligence increases resilience, allowing individuals to navigate stress and setbacks with greater ease. This is particularly important when dealing with oxytocin blockers triggered by external pressures or trauma. By building emotional resilience, individuals are better equipped to restore oxytocin balance and maintain healthy relationships.

3. **Leadership and Social Influence**: Emotional intelligence is a key trait of effective leadership. Leaders with high emotional intelligence are more adept at understanding their team members' emotional needs and creating an environment of trust and cooperation. This is especially important in professional settings where oxytocin blockers can disrupt workplace dynamics and hinder collaboration.

Conclusion

Building emotional intelligence is a powerful way to combat the effects of oxytocin blockers and improve human connection. By cultivating self-awareness, empathy, emotional regulation, and resilience, individuals can navigate the challenges posed by oxytocin blockers while fostering stronger relationships and a more compassionate society. The tools and strategies outlined in this chapter provide a roadmap for enhancing emotional intelligence and mastering the art of connection, even in the face of obstacles. In a world that increasingly challenges our ability to bond, emotional intelligence becomes our most valuable asset in creating deep, meaningful connections and overcoming the barriers to true human connection.

Chapter 22: Reversing the Effects of Oxytocin Blockers

Oxytocin blockers, whether stemming from stress, trauma, or the modern challenges of social disconnection, can profoundly impact our emotional and social well-being. The good news is that, like many biological processes, the effects of oxytocin blockers can be reversed. By understanding the neurochemistry behind these blockers and applying targeted strategies, it is possible to restore the balance of oxytocin and reclaim the deep human connections that are vital to our emotional and psychological health. This chapter will explore the most effective techniques for reversing the effects of oxytocin blockers, ranging from psychotherapy to lifestyle changes and behavioral interventions.

Understanding the Neurochemistry of Oxytocin Blockers

Before diving into the methods for reversing oxytocin blockers, it's essential to understand the underlying biology. Oxytocin is a powerful hormone that influences our social behavior, emotional regulation, and bonding capabilities. When oxytocin is blocked—either due to trauma, stress, or environmental factors—the brain's natural pathways for connection and empathy become hindered.

Oxytocin blockers may alter the functioning of certain receptors, such as the oxytocin receptor (OXTR), which is primarily involved in feelings of trust, affection, and social bonding. This disruption can lead to emotional numbness, difficulty forming connections, and a decreased capacity for empathy. The good news is that the brain is neuroplastic, meaning it has the ability to reorganize itself and form new neural connections—allowing for the restoration of oxytocin levels and the re-establishment of healthy social behaviors.

Psychotherapy: Addressing the Root Causes

One of the most effective ways to reverse the effects of oxytocin blockers is through psychotherapy. Psychotherapy offers a safe space to explore and process the underlying emotional, psychological, or social factors that may be contributing to the decrease in oxytocin levels.

1. **Trauma-Informed Therapy**

   Trauma, whether it stems from childhood experiences, relationship difficulties, or major life events, is one of the leading causes of oxytocin depletion. Trauma can severely affect the body's natural oxytocin production, particularly when the person feels unsafe, unsupported, or disconnected. Trauma-informed therapies—such as EMDR (Eye Movement Desensitization and Reprocessing) or somatic experiencing—help to rewire the brain and body's responses to stress, ultimately restoring oxytocin pathways.

2. **Attachment-Based Therapy**

   Since oxytocin plays a critical role in the formation of attachment bonds, attachment-based therapies focus on repairing disrupted or insecure bonds between individuals. These therapies emphasize fostering safe, empathetic, and supportive relationships, allowing individuals to re-experience healthy connection and, in turn, promote the natural release of oxytocin.

3. **Cognitive Behavioral Therapy (CBT)**

   Cognitive Behavioral Therapy (CBT) can be an effective method for reversing oxytocin blockers, especially in cases where negative thought patterns contribute to social withdrawal or emotional detachment. By identifying and challenging harmful beliefs—such as "I am unlovable" or "People can't be trusted"—CBT helps individuals reframe their perception of themselves and others, restoring the emotional balance necessary for healthy connection.

4. **Social Support and Interpersonal Therapy**

Often, oxytocin blockers arise from social isolation or a lack of meaningful relationships. Interpersonal therapy (IPT) emphasizes the importance of social support and communication in maintaining mental and emotional well-being. By improving communication skills and fostering supportive relationships, individuals can cultivate a network that actively promotes the release of oxytocin.

Neuroplasticity: Rewiring the Brain for Connection

As mentioned earlier, the brain's neuroplasticity is a key factor in reversing the effects of oxytocin blockers. Neuroplasticity refers to the brain's ability to reorganize itself by forming new neural connections. By engaging in certain activities, individuals can rewire their brain to be more receptive to oxytocin and strengthen the neural pathways involved in empathy, trust, and social bonding.

Mindfulness and Meditation

Practical Exercise

Repetitive Positive Social Interactions

Practical Exercise

Behavioral Changes: Rebuilding Connection Through Action

In addition to psychotherapy and neuroplasticity exercises, certain behavioral changes can directly help reverse the effects of oxytocin blockers. These changes focus on creating environments that naturally foster oxytocin production and cultivating habits that promote emotional connection.

1. **Physical Touch and Affection**

   Physical touch is one of the most effective ways to increase oxytocin levels. Hugging, cuddling, or even casual touch such as hand-holding or a pat on the back can stimulate oxytocin production and promote feelings of trust and connection. For individuals who have experienced emotional or physical trauma, re-establishing safe and consensual physical touch is a crucial step in reversing oxytocin blockers.

2. **Engaging in Acts of Kindness**

   Acts of kindness, whether big or small, activate the brain's oxytocin system and promote feelings of emotional warmth. Helping others, practicing generosity, and expressing gratitude are all behaviors that can stimulate oxytocin production. In addition to benefiting the recipient, these acts also reinforce the giver's emotional well-being and sense of connection.

3. **Healthy Lifestyle Choices**

   Diet and exercise also play a role in reversing the effects of oxytocin blockers. Regular exercise, especially activities that involve social interaction (such as group sports or fitness classes), can stimulate oxytocin production. Additionally, eating a balanced diet rich in omega-3 fatty acids, antioxidants, and nutrients that support brain health can help maintain optimal oxytocin function.

4. Creating a Safe and Nurturing Environment

Finally, creating an environment that supports emotional well-being is critical for reversing oxytocin blockers. This means prioritizing safe, supportive relationships, reducing stressors where possible, and fostering an environment that promotes emotional expression and connection. Whether at home, at work, or in social settings, the more nurturing and empathic the environment, the more likely it is that oxytocin will flow freely.

Conclusion

Reversing the effects of oxytocin blockers is not a quick or simple process, but with the right tools and strategies, it is entirely possible. Through psychotherapy, neuroplasticity exercises, positive social interactions, and intentional behavioral changes, individuals can restore the balance of oxytocin in their bodies and minds. By taking proactive steps to overcome the negative effects of these blockers, we can re-establish the bonds of trust, love, and empathy that are fundamental to our well-being. Ultimately, the goal is to create a life in which oxytocin flows freely, fostering a deeper connection with others and a richer, more fulfilling emotional experience.

Chapter 23: Cultivating Compassion and Empathy in a World of Oxytocin Blockers

In a world where stress, trauma, and technological distractions are commonplace, cultivating compassion and empathy is more crucial than ever. These two emotions are at the heart of human connection and social cohesion, and yet, the rise of oxytocin blockers—whether natural or environmental—often hinders their full expression. By understanding the role of oxytocin in fostering compassionate behavior, we can take concrete steps to nurture both our own empathy and that of others, even in a society that increasingly seems to prioritize individualism and disconnection.

The Role of Empathy in Human Connection

Empathy—the ability to understand and share the feelings of another—requires more than just cognitive awareness; it is deeply tied to emotional resonance, and this emotional connection is driven by oxytocin. Oxytocin, often referred to as the "bonding hormone" or the "love hormone," plays a vital role in the neural circuits responsible for empathy. When oxytocin levels are balanced and receptors in the brain are activated, we naturally feel drawn to help others, to listen with care, and to respond with compassion. This biological foundation for empathy is essential for forming bonds, whether in romantic relationships, friendships, or within communities.

However, oxytocin blockers—whether the result of stress, trauma, or societal factors—can impair this emotional resonance, making it harder for individuals to empathize with others. In some cases, people may even become detached from their own emotions, which, in turn, inhibits their ability to connect with the emotions of others. The more we recognize the neural and hormonal underpinnings of empathy, the better equipped we become to foster these qualities, both within ourselves and in the wider world.

How Oxytocin Blockers Hinder Compassionate Behavior

The challenges of modern life contribute significantly to the increase of oxytocin blockers. The rise of technology and social media, for instance, often fosters isolation rather than connection, creating environments where individuals feel more disconnected, less attuned to others' emotional needs, and less willing to engage in acts of kindness or empathy. Social media, while facilitating virtual connections, can also create an environment that magnifies comparison, envy, and competition—emotions that trigger the release of stress hormones like cortisol, which, in turn, can block oxytocin production.

Further, trauma, whether from past abuse, neglect, or emotional wounds, can create a psychological defense mechanism where the individual closes off their ability to feel or express empathy. When the brain's oxytocin pathways are blocked by trauma or prolonged emotional pain, it can lead to a state of emotional numbness or indifference, making it more difficult for individuals to engage with others in a meaningful way.

The global rise in mental health challenges, such as anxiety, depression, and PTSD, has also contributed to this disconnection. These conditions can be exacerbated by oxytocin blockers, which make it harder for individuals to form the social bonds necessary for healing and recovery.

Reconnecting with Compassion in a Disconnected Society

Despite these challenges, there are numerous strategies for overcoming oxytocin blockers and fostering compassion and empathy in today's society. Reversing the effects of oxytocin blockers requires a concerted effort to create environments—both within ourselves and in our relationships—that prioritize connection, understanding, and support.

Fostering Vulnerability and Openness

Practical Exercise

Engaging in Acts of Kindness

Practical Exercise

Creating Empathetic Communities

Practical Exercise

Mindfulness and Compassion Meditation

Practical Exercise

Reducing Stress and Prioritizing Self-Care

Practical Exercise

The Power of Connection: How Small Efforts Can Lead to Big Changes

In a world where oxytocin blockers are becoming increasingly prevalent, even small efforts to cultivate compassion and empathy can have profound effects on individuals and communities. By practicing vulnerability, engaging in acts of kindness, creating empathetic environments, and prioritizing mindfulness and self-care, we can reverse the negative effects of oxytocin blockers and help build a more compassionate, connected world.

Ultimately, the ability to connect with others in meaningful, compassionate ways is one of the greatest gifts we can offer. As we unlock the science of oxytocin and actively cultivate empathy, we not only improve our own well-being but also contribute to the collective healing and growth of society. In doing so, we can ensure that empathy— along with the transformative power of oxytocin—becomes a force that binds us together rather than something that gets blocked in the face of adversity.

Chapter 24: Living a Life Free from Oxytocin Blockers

In our increasingly fast-paced, disconnected world, the concept of living a life free from oxytocin blockers may seem like an impossible ideal. However, it is both attainable and transformative. Mastering oxytocin blockers is not just about restoring a hormone balance in the brain—it's about actively creating environments and practices that prioritize connection, well-being, and emotional resilience. In this chapter, we explore the practical ways to cultivate a life where oxytocin flows freely, allowing for deeper relationships, improved mental health, and a more compassionate world.

How to Create an Environment that Promotes Oxytocin

The first step in living a life free from oxytocin blockers is to create environments that actively promote the release of oxytocin. These environments, whether physical, social, or emotional, must foster connection, safety, and understanding. The spaces we inhabit —both externally and internally—play a crucial role in either stimulating or blocking oxytocin production.

1. Fostering Connection in Relationships

Human connection is the primary trigger for oxytocin release. Building and nurturing healthy, trusting relationships with family, friends, colleagues, and romantic partners creates an emotional foundation that facilitates oxytocin production. Regular, meaningful interactions—whether through active listening, physical touch, or shared experiences—are essential for creating bonds that support both emotional and physiological well-being.

Practical Strategy

2. Creating Safe and Supportive Social Environments

Our social environment directly influences our ability to form meaningful bonds and experience empathy. Workplaces, community spaces, and even social media platforms can either foster or inhibit oxytocin release, depending on the level of emotional safety, trust, and openness they offer.

Practical Strategy

3. Prioritizing Positive Physical Touch

Physical touch is one of the most powerful ways to stimulate oxytocin release. Simple, affectionate gestures such as hugging, holding hands, or even a pat on the back trigger oxytocin production and foster feelings of connection. In fact, studies show that even brief moments of physical contact with a loved one can significantly improve mood and emotional well-being.

Personal Growth and Emotional Well-Being Strategies

Living a life free from oxytocin blockers requires us to engage in personal development practices that enhance emotional regulation, resilience, and self-awareness. This means taking proactive steps to manage stress, heal emotional wounds, and cultivate habits that promote emotional well-being.

1. Building Emotional Intelligence (EQ)

Emotional intelligence (EQ)—the ability to recognize, understand, and manage our own emotions, as well as the emotions of others—is essential for cultivating compassion and empathy. People with high EQ are better at navigating interpersonal relationships, responding to others' needs, and creating emotionally supportive environments. A well-developed EQ also enables us to prevent or manage the blockers that arise from negative emotions like anger, fear, or jealousy.

Practical Strategy

2. Cultivating Self-Compassion

In addition to compassion for others, self-compassion is a critical element of living a life free from oxytocin blockers. Negative self-talk, guilt, and shame can all contribute to emotional distress and the release of stress hormones that interfere with oxytocin production. Cultivating self-compassion involves treating ourselves with the same kindness, understanding, and patience that we would offer a friend in times of difficulty.

Practical Strategy

3. Managing Stress and Practicing Relaxation

Chronic stress is one of the most common blockers of oxytocin. Stress hormones like cortisol can suppress oxytocin production, making it harder for individuals to connect emotionally with others. Learning to manage stress is crucial for maintaining a healthy flow of oxytocin in the body. This involves both proactive stress-reduction techniques and the development of coping mechanisms that prevent stress from escalating into chronic conditions.

Practical Strategy

The Long-Term Benefits of Overcoming Oxytocin Blockers

Living a life free from oxytocin blockers not only enhances emotional connection but also promotes long-term mental, physical, and social well-being. The effects of consistently fostering oxytocin release through positive relationships, personal growth, and self-care can have a profound impact on our overall quality of life.

1. Enhanced Physical Health

Oxytocin is known to have a wide range of health benefits, from reducing inflammation and promoting heart health to improving immune function and reducing pain perception. A life with regular oxytocin release can help reduce the risk of chronic diseases associated with stress, such as cardiovascular disease, diabetes, and autoimmune disorders. By prioritizing connection and emotional well-being, we also foster physical vitality.

2. Improved Mental Health

Oxytocin is a natural antidepressant, as it promotes feelings of calm and well-being. Regular oxytocin release has been shown to reduce symptoms of anxiety, depression, and PTSD. Furthermore, fostering emotional connection through compassionate relationships helps to buffer the impact of emotional trauma, reducing the likelihood of experiencing long-term mental health challenges. The more we work to nurture and protect our oxytocin levels, the more resilient we become in the face of life's inevitable difficulties.

3. Stronger Social Bonds and Communities

As oxytocin enhances social bonding, its consistent presence in our lives also strengthens the communities we are part of. By cultivating an environment where oxytocin is encouraged—through both our personal interactions and the larger social structures we contribute to—we can foster a sense of collective well-being. Stronger communities are not only more supportive and resilient but also more compassionate and cohesive.

4. A More Compassionate World

Ultimately, living a life free from oxytocin blockers contributes to a broader societal transformation. The more individuals prioritize oxytocin-promoting practices, the more they contribute to a global culture of empathy, understanding, and compassion. As we collectively learn to manage and reduce oxytocin blockers in our lives, we create a ripple effect that can inspire a more connected and supportive world.

Conclusion

Living a life free from oxytocin blockers is an ongoing process that requires intention, practice, and patience. By fostering environments that promote emotional connection, investing in personal growth, and making self-care a priority, we can create lives that are rich in empathy, resilience, and compassion. The long-term benefits of overcoming oxytocin blockers extend far beyond individual well-being; they pave the way for stronger relationships, healthier communities, and a more connected world. As we master the art of cultivating oxytocin, we unlock the full potential of human connection, enriching our lives and the lives of those around us.

Chapter 25: Conclusion: Mastering Oxytocin Blockers for a Healthier, Connected Life

Throughout this journey, we've explored the profound influence that oxytocin, the so-called "love hormone," has on our lives. We've examined how it shapes our biology, guides our behavior, and sustains our social bonds. But just as oxytocin has the power to nurture connection and well-being, so too do its blockers—stress, trauma, environmental toxins, and even modern lifestyle factors—work to create barriers between us and our fullest potential for human connection and emotional health.

Mastering oxytocin blockers is not just about learning how to mitigate their effects. It's about cultivating a life where oxytocin flows freely, supporting deeper relationships, mental and physical health, and a more compassionate society. It's a path to emotional resilience, healing, and profound personal transformation. Let's now recap some of the key takeaways from this book and outline how you can continue to cultivate a life filled with connection, empathy, and well-being.

The Power of Oxytocin in Human Connection

Oxytocin is a powerful neuropeptide that plays a critical role in bonding, trust, and social cohesion. From the first moments of life, it facilitates the parent-child bond, strengthens romantic relationships, and nurtures friendships. It is essential for fostering empathy, compassion, and emotional regulation. Oxytocin's ability to mitigate stress, promote healing, and even improve physical health cannot be overstated. As we've seen throughout this book, by promoting the release of oxytocin, we can improve our relationships and strengthen our emotional and mental health.

Understanding Oxytocin Blockers

While oxytocin's benefits are clear, it's equally important to understand the various blockers that can disrupt this process. These blockers—whether caused by stress, fear, isolation, trauma, or even the fast-paced nature of modern life—pose significant challenges to our ability to connect and thrive. For many, these blockers are subtle and pervasive, gradually eroding our capacity for emotional closeness and creating a sense of disconnection from others. However, understanding how they work, and how they manifest in our bodies and minds, is the first step toward overcoming them.

Practical Strategies for Mastery

The good news is that mastering oxytocin blockers is not a matter of passive acceptance but active engagement. Throughout the chapters, we have outlined a variety of strategies to help you reverse the effects of these blockers and promote a life filled with connection and emotional health. These strategies include:

- **Building and nurturing relationships** that promote emotional safety and trust.

- **Creating environments** that encourage connection and empathy, both in personal and professional settings.

- **Adopting healthy lifestyle practices** such as balanced nutrition, regular exercise, and adequate rest that support oxytocin production.

- **Engaging in mindfulness and meditation** to regulate emotions and reduce the impact of stress and anxiety.

- **Therapeutic approaches** like psychotherapy, trauma recovery, and interventions designed to repair disrupted bonds.

- **Developing emotional intelligence** to navigate conflicts, enhance empathy, and foster deeper social connections.

Each of these tools is a step toward freeing yourself from the constraints of oxytocin blockers, allowing you to lead a more connected, fulfilling life.

The Ripple Effect of Connection

As we work to master oxytocin blockers on an individual level, the positive effects ripple outward. By creating a life that prioritizes connection, emotional health, and empathy, we become catalysts for change in our families, workplaces, and communities. As more individuals and groups cultivate compassion and understanding, we collectively build a more connected, resilient society. In this way, the mastery of oxytocin blockers is not just about improving one's personal well-being—it's about fostering a world where human connection is the cornerstone of society.

A Roadmap for the Future

To truly master oxytocin blockers, it requires ongoing effort and commitment. This is not a one-time fix, but a lifelong process of personal growth, emotional awareness, and conscious connection. Below is a roadmap for integrating these principles into your life:

1. **Start with Self-Awareness**: The first step in overcoming oxytocin blockers is to recognize when they are present. Pay attention to your emotional triggers, your physical responses, and the impact of your environment on your mood and behavior. Being self-aware allows you to take proactive steps to counteract blockers and promote connection.

2. **Cultivate Connection Every Day**: Make connection a priority in your life. Whether it's through a hug, a kind word, or a shared experience, aim to foster meaningful interactions with the people around you. Start small and gradually build a network of supportive, empathetic relationships.

3. **Build Emotional Resilience**: Engage in practices that help you regulate your emotions and reduce stress. Whether it's through exercise, mindfulness, or therapy, prioritize your emotional well-being as part of your daily routine. Remember, reducing the impact of blockers is a continuous process.

4. **Create Positive Environments**: Make an effort to surround yourself with spaces—both physical and social—that encourage connection. This could mean setting boundaries around technology use, fostering supportive communities, or engaging in practices that promote group cohesion, such as team-building or group therapy.

5. **Continue to Learn and Grow**: As you work toward mastering oxytocin blockers, remain open to new research, tools, and techniques that can support your journey. Science and self-development are ever-evolving fields, and new breakthroughs in neuroscience, therapy, and emotional intelligence will continue to offer additional avenues for growth.

Final Thoughts: A Connected Future

Mastering oxytocin blockers is a transformative journey—one that requires both intention and effort. But the rewards are well worth it. By fostering deeper emotional connections, embracing compassionate relationships, and taking proactive steps to regulate our biological and emotional responses, we create a world where empathy, understanding, and well-being thrive.

In a time when disconnection is increasingly common, the ability to unlock the power of oxytocin and overcome its blockers is both a personal and collective responsibility. By nurturing connection in ourselves and others, we can create a healthier, more compassionate world for generations to come.

As you move forward, remember: every small step you take toward greater connection, every act of kindness and understanding, contributes to the larger goal of building a world where oxytocin flows freely, where love and empathy are the driving forces of human interaction. Mastering oxytocin blockers is not just a matter of individual well-being—it is a revolutionary act of creating a more connected and compassionate society.

The path to mastery is in your hands. Embrace it with openness, intention, and a commitment to living a life rich with human connection, emotional resilience, and profound well-being.